THE COMING FOOD CRISIS

THE COMING FOOD CRISIS

FRANK FORD

with Jamie Buckingham

Chosen Books
Lincoln, Virginia 22078

All Scriptures are taken from the King James Version of the Bible unless otherwise identified as follows:

Library of Congress Cataloging in Publication Data
Ford, Frank.
The coming food crisis.
1. Food supply. 2. Food-Preservation.
3. Agriculture. I. Buckingham, Jamie.
II. Title.
TX357.F68 363.8 82-1286
ISBN 0-912376-78-3 AACR2

Published by Chosen Books Publishing Co., Ltd.

Printed in the United States of America

Dedication and Acknowledgment

This book is dedicated to my dear friend, Doug Chatham, whom God used to make it a reality. Doug's excellent research, his extensive knowledge in the field of food preparation and preservation, and his hard work on the first draft of the manuscript make this more than just a readable and insightful book. It will also be a valuable addition to your permanent library, providing some practical references for the days to come.

Doug was tough before God saved his soul. He is now tough in the way that Jesus taught and demonstrated in His pattern life.

Frank Ford
Arrowhead Mills
Hereford, Texas

Preface

Every time I walk through our storage room on my way back to my writing studio, I pass a wall stacked ceiling-high with boxes labeled "The Simpler Life." The boxes are filled with things like Dehydrated Apple Dices, Freeze-Dried Potatoes, Nonfat Dry Milk, Nitrogen-Packed Whole-Grain Wheat, Nitrogen-Packed Banana Flakes, and lots of other good things. There is enough food in those boxes to provide healthful nutritional meals for my wife and myself for an entire year.

I am not the only one storing food for the coming food crisis. Everywhere I go around the nation I find families doing the same thing. None of them seems to know why. They are, it seems, like animals in the forest who somehow have heard an unspoken cry that it is time to head for the ark.

We're doing other things to get ready for the lean years. We're gardening, raising livestock, and rearranging our lives so we can quickly adapt when things get tight in America.

Like Frank Ford, I do not feel this is "the end." Rather, I believe we are about to enter a crisis much like that experienced in the time of Joseph in Egypt—a crisis brought on by greed, poor management, and a humanistic government which will send us hurtling into a great time of physical want.

When I first began to hear that we should store food, I was skeptical. In fact, I was defensive.

America? Go hungry? Nonsense! That was for underdeveloped nations such as Bangladesh, Cambodia, Somalia, and Biafra. I scoffed at the idea of hard times ahead.

I no longer feel that way.

Frank Ford is a man of God, a man fulfilling the true spiritual prophetic office, who has the credentials to make him one of the nation's leading experts on food supplies. So when Frank Ford says I ought to be storing food, I listen.

But he says far more than that. He says the coming food crisis is for more than preparing; it will be a time of opportunity. For the Christian in particular, this will be a great time of caring and sharing.

There is no place in the Christian ethic for the bombshelter mentality where I defend my food supply with guns. Rather, Ford says, we should store in order to help those who cannot (or will not) store for themselves.

This book is not another doomsday book. It is not even a book about end times, although what Ford discusses in these pages are certainly signs of end times. Rather, it is a book of hope and opportunity. It is a straight-from-the-shoulder, highly controversial book about the way God is dealing with our nation, what we can expect to happen (and why), and how we can prepare so we might not just live through the crisis—but live abundantly.

Jamie Buckingham
Melbourne, Florida

Introduction

In January, 1973, Frank Ford turned forty. He had then what appeared to be the essence of the American Dream. In college, in the armed services, and later in his community and profession, he had been accorded high honors and given many responsibilities. With a fine family, a bumper wheat crop in prospect, and his natural-food business prospering for the first time after a dozen difficult years, all was well—except in the most important realm of all.

The problems of the world had finally begun to stagger the strong will of Frank Ford. Things were getting worse in spite of his best efforts. A period of four months of sleeplessness and instability carried Frank beyond his tenacity, his perseverance, and his own strength. In April, a once-proud man who needed no help bowed his knee and asked for it. Jesus Christ came into his life.

"The gift of the Holy Spirit," a phrase Frank had never heard before, began to ring in his inner spiritual ear, and one month later, the Lord graciously and lovingly poured the streams of "Living Water" over His new believer. Renewed and hungry for God's Word, Frank devoured first the New Testament and then the entire Bible in the next few weeks, reading and underlining five or six hours a day.

Filled with joy and excitement, he saw miracles as many accepted the free gift of grace in Jesus Christ. Street ministry in Berkeley, California, Austin, Texas, and in his own hometown, and a genuine love for people, filled his cup to overflowing. He gave his testimony as many doors opened. He taught the Word of God and let the Holy Spirit use him for words of knowledge, healing, and prophecy. The former artillery captain was now on the front line for Jesus Christ, and loving it. But then came new directions. God spoke to Frank and said, "I have many who can preach the Gospel better than you. I want you back in your business."

Frank had already been struggling with an idea about helping people set up food reserves to prepare for harder times. Returning to the business he had founded, Arrowhead Mills, Frank soon developed Simpler Life Reserve Foods. It was an idea whose time had come. The storage-foods aspect of his business began to be blessed of the Lord.

In the years that followed, Frank Ford's name became well-known in the natural-foods and reserve-foods industries. Through his speaking and writing, he has become a recognized authority on the trends and economics of food production. Frank also has acquired a thorough knowledge of biblical prophecy; as a result, he has been gifted with the ability to discern what will happen next. I think he has been called of God to be one of the "watchmen on the wall" who sound forth warnings and approaching danger. I predict his voice will be heard more and more in the days of deepening crisis.

God has made Frank Ford like one of the biblical "sons of Issachar" of I Chronicles 12:32. They, you may recall, "had understanding of the times, to know what Israel ought to do. . . ." If ever there was a time when the people of God needed "sons of Issachar," it is now. Knowing Frank, which is my privilege, is a little like knowing a son of Issachar, a Joseph and a David rolled into one.

While this book deals with an impending crisis, it is basically a positive book with practical help and loving assurances to those who have put their trust in the Lord Jesus Christ. It contains information that we need now and will need in the days to come. I suspect you will find yourself rereading it often, and enthusiastically sharing it with your friends.

Harald Bredesen
Escondido, California

Contents

1
Signs of Crisis

Our world is about to enter into the most severe food crisis in recorded history. This will be a crisis brought about by our own stupidity, selfishness, greed, and failure to live under the laws and principles of God.

The food crisis of the 1980s will be even worse than the oil crisis of the 1970s, and it will mean starvation for many. It will cause riots in cities and necessitate a radical change in life-style for nearly everyone. Governments will topple. Revolutions will take place. The entire world economy will be shaken apart and put together again.

Yet I see great hope in this coming crisis, for our world needs to be shaken apart and put back together again—politically, economically, and spiritually.

A number of us who work in the area of food production have known for several years such a crisis was coming. For various reasons, however, few have been willing to say much about it. Now I feel forced to stick my neck out and become a prophet. Not a prophet of doom, but a prophet of crisis—and yet of hope.

My purpose in writing this book is not to scare you into thinking all is lost, but in hopes you will do something to prepare, to care, and to share as the food in this nation runs out.

Remember that what I am telling you is not just my opinion, but it is also the studied opinion of many food specialists all over the world who are predicting lean years ahead.

Do you remember the story of Joseph in the Bible? He followed God's direction when forewarned of the seven years of famine about to come. The people listened to him and made preparations; as a result, ancient Egypt and the children of Israel were spared from sure destruction.

Now, our nation is faced with a similar situation. We have been living in years of plenty, but the years of want are almost upon us. It is time to prepare for the future. We have no more years to sit back and wait.

Our world has changed dramatically in just the past twenty-four months. The decade of the eighties has already impacted human history with almost unparalleled change. Futurists attempt to build what are called "scenarios" to describe the days ahead. Biblical scholars see an approaching tribulation followed by a glorious millennium. One fact is certain. Any secure ties we may have had with a familiar world order as we have known it are being permanently severed by current events.

Consider how profoundly our world has been affected by recent events in Afghanistan, Poland, Iran, Africa, and Central America. Think of the implications of the successful tests of the space shuttle and laser weaponry. What does the Middle East race for atomic power mean for the rest of the world? Think of the changes brought by inflation. Think of the effect that the changes in our weather patterns are already having on food production. The failure of "detente" and the renewal of the arms race with Russia, the possible climatological and geological effects of increasing volcanic and earthquake activity, rising unemployment, and the faltering dollar . . . all are making the future uncertain. And all are working to bring us a worldwide food shortage.

It is no longer a matter of "if" but "how soon?"

As I see it, reasons for the present crisis fall into six major categories:

1. *The Shrinking World Food Supply.*

World food supplies are shrinking at an alarming rate. In every major food-producing nation but the United States, there is an indication that grain carry-overs are shrinking every year. ("Carry-over" is the term used for how much of the old crop is left when the new crop comes in.) Weather conditions have repeatedly devastated the huge Russian croplands and other key food-growing areas.

According to a November 9, 1981, article in *Newsweek* magazine, the 1981 Russian grain harvest was critically short for the third straight year. The Russians are having to import millions of tons of wheat and other grains from foreign suppliers, including the United States. However, the problems of their failing agricultural system are still not solved, and we may expect their dependence on foreign suppliers to increase.

In the same article, *Newsweek* reported, "Food shortages are a way of life, even in the Soviet Union's farmlands." Meat has largely disappeared from Russian state-run food stores since the 1960s. Milk is now sold by prescription only, and that just for infants: one liter a day for babies up to a year old, half a liter for those over one and up to two. Butter and cheese are almost never available.

United States food production has been hit hard by weather that has been either too dry or too wet. In addition, 400 insect species have developed partial or total immunity to pesticides, with catastrophic results. In the last thirty years, harmful insect populations have doubled, despite the fact that pesticide spraying has been increased twelvefold. In 1981, army worms and corn borers appeared in the corn belt in record numbers. Fruitflies all but ruined California's fruit exporting. If half the funds which have been spent on chemicals had been spent on research for biological and predator control of insect pests, we would be in much better shape in the world of entomology. But ladybugs, lacewing flies, and other insects that are beneficial to crop production do not give grants to colleges and universities. Chemical companies do.

It is significant to many of us who believe in God that one of the manifestations of God's wrath against His people was increased destruction of crops by insects:

> Thou shalt carry much seed out into the field, and shalt gather but little in; for the locust shall consume it. Thou shalt plant vineyards, and dress them, but shalt neither drink of the wine, nor gather the grapes; for the worms shall eat them (Deuteronomy 28:38, 39).

2. *The Loss of Topsoil.*

Adequate topsoil is very important to the production of good crops, yet about one-third of our approximately 400 million acres of cropland is severely eroded, with many areas almost denuded of topsoil.

According to a March, 1981, article in the *Smithsonian Magazine,* the United States is on the verge of returning to the dust-bowl days of the 1930s, when wind and water ravaged 282 million acres of American farmland. In some ways, soil erosion is worse today than in dust-bowl days. In 1980, 5.1 million acres of Great Plains land were damaged by routine wind erosion, almost twice as many acres as were damaged the previous year.

Nationwide, water erosion is taking an even greater toll. Much of our rich topsoil, laden with chemical fertilizers and pesticides, ends up in streams, rivers, and lakes, polluting these waterways for recreation and water supply, and destroying fish and wildlife. Agricultural chemicals have become the single biggest pollutant in two-thirds of the river basins in the United States.

According to the *Smithsonian* article, the topsoil upon which crops can best be grown varies in depth from a few inches to several feet. As a general rule, an acre of land can lose five tons of soil per year to erosion without permanent damage because under ideal conditions that amount is matched by the formation of new topsoil through natural processes. (This so-called "tolerance value" of five tons may sound enormous, but spread out over an acre it amounts to a layer only a fraction of an inch in depth.)

What scares experts today is that the average erosion figure for *all* United States cropland is already reaching the five-ton tolerance value—while in some vital farming areas, essentially all the topsoil is already gone. A recent Department of Agriculture survey warns that at current erosion rates, corn and soybean yields in the corn belt states may drop by as much as thirty percent over the next fifty years.

Some areas are being devastated by erosion rates much higher than the national average, however, and their crop production is dropping much faster. In the western Tennessee counties bordering the Mississippi River, erosion now averages thirty to forty tons per acre each year. Some farms in that area are registering soil losses of

150 tons per acre. Iowa produces twenty percent of the entire United States corn crop and fifteen percent of its soybean harvest. When Iowa first was plowed and planted about 100 years ago, it was covered with rich, black topsoil to an average depth of sixteen inches. Now only about eight inches remain. Intensive high-technology farming methods have taken their toll.

A very rapid reduction in land fertility is also occurring. Mineral content of topsoil in many areas has dropped severely in the last ten years. Chuck Walters, publisher of *Acres, USA*, has often stated his opinion that because of the decline of mineral ingredients in food crops, the "mental acuity" of Americans has diminished in the past decade. The removal of *life* from our food supply by multinational agribusiness and food-refining corporations has contributed to this growing problem.

In a 1981 interview published in *U.S. News & World Report*, Lester R. Brown, a former Agricultural Department official who now heads the Worldwatch Institute, a research organization analyzing global problems, says, "There is a very real possibility that a global food crisis may unfold during the 1980s, much as the energy crisis did during the 1970s. . . . Just as Middle Eastern oil reserves are being depleted, so too are North American soils being depleted. We are, in effect, mining our soils—an agronomic form of deficit financing—in order to satisfy current demand."

Brown goes on to say, "The world already has more than 100 countries that cannot produce enough food to feed themselves. Unless we in North America act soon to curb the loss of precious topsoil, we all will face an immeasurably greater and potentially far more catastrophic problem ten to twenty years down the road."

My own contention is that we don't have ten years, much less twenty years. The crisis is here. In a few short years, the demand for food will far outdistance our ability to produce it.

Brown continues: "The farmer who is losing topsoil at an excessive rate has two choices: go bankrupt in the short run, or go bankrupt in the long run."

In the same interview, the former government offical makes this startling statement: "The easy production gains through hybridization of corn and intensive use of chemical fertilizers are now largely behind us. Twenty years ago, another pound of nitrogen fertilizer would get you another twenty pounds of corn. Today another pound of fertilizer will get you only six or eight pounds of additional corn.

"The hard fact is that the outer limits of what U.S. agriculture can produce are determined by the amount of incident sunlight on the land. We are boxed in by the basic biochemistry of photosynthesis."

Soil is biology. It is made up not only of chemical elements but also of micro-life which turns organic matter into rich humus and gives the wonderful aroma to a handful of healthy soil when you start your garden in the spring. The balance of nature which God ordained in creation is a cycle of life which renews itself. Where there has been stewardship of the soil that seeks to maintain this balance, good topsoil has been built up rather than destroyed.

Unfortunately, man in his ignorance and insensitivity has more often mined rather than farmed the soil. After World War II there was a need for new markets for the nitrates, poisons, and other new chemical compounds which had been utilized as engines of death. Agriculture became the victim. In recent years, pesticides and herbicides used in American agriculture have not kept pace with the increasing resistance of insects and of weed pests. A vicious cycle of death has replaced the intended cycle of life.

It's not all bad news here. Some farmers have turned away from "business as usual" and have become strong proponents of working with, rather than against, nature. They now farm "biologically," helping the life forces in the soil to restore themselves wherever possible.

One example of biological farming is what we are doing in Deaf Smith County, Texas, an area that has long been known for the unusually high fertility of its soil and the valuable nutrients carried in its water. We ask farmers who sell to us to use organic methods, and we routinely conduct inspections to ensure they are doing it.

Thousands of acres in this area have been covered with biodynamic compost, using the abundant supply of manure from the many feeding-yard operations that are located in Hereford, Texas. Lab tests over the years have continually shown that grains grown here are considerably above the national average in protein content. Tests also show that essential trace minerals are abundantly present in foods produced in Deaf Smith County.

Beneficial insects such as trichogramma wasps, ladybugs, lacewing flies, and praying mantises are used to eliminate the need for dangerous chemical pesticides. What we are doing here is only a small example of the type of farming that must be done nationwide to rebuild tired soil and produce more wholesome food.

3. *International Politics.*

The third problem that affects our food situation is the political hostility that exists today among so many nations. Right now there is a race between Russia and the Western nations for control of Middle East oil. The most important waterway in the world is the narrow Strait of Hormuz that links the Persian Gulf with the Arabian Sea. The passage through this strait narrows at one point to only a few miles. Through that narrow passage, bordered on one side by Iran and on the other side by the tiny nations of South Yemen and Oman, passes daily thirty percent of the oil imported by America, sixty-three percent imported by Western Europe, and seventy-three percent imported by Japan.

Without that oil from the Persian Gulf, America's economy would be severely shaken. Even a brief interruption of the thin line of tankers stretched halfway around the world would result in oil rationing in the U.S.

Many observers predict a Soviet takeover very soon in unstable Iran, like that already accomplished in Afghanistan. It seems likely that Russia will continue her aggression to gain control of the Persian Gulf. Already she has signed a friendship treaty with South Yemen, and since then has armed that small nation with sophisticated fighter planes and formidable T-54 tanks. MIGs believed to be piloted by Cuban pilots are on regular patrol of the sea lanes in the Persian Gulf, the Strait of Hormuz, and the Arabian Sea. Guerrilla warfare is being waged from South Yemen against tiny Oman. It is conceivable that Russia could soon have complete control of the Strait of Hormuz. If this were to happen, the U.S. would face the loss of nearly one-third of its total oil supply, upon which our agriculture is heavily dependent.

Several years ago the U.S. grain embargo against Russia—in

retaliation for the Soviet invasion of Afghanistan—greatly harmed our domestic grain market. Interestingly, the invasion was partly accomplished with vehicles produced at the Kama River truck plant, financed and built by U.S. companies. The embargo punished some American producers and taxpayers, while leaving the flow of high technology from other U.S. businesses untouched. The U.S. sale of computers and other technology could help the Soviets place every United States city under the threat of massive nuclear missile attack.

The Soviets invaded Afghanistan to obtain the minerals and natural gas of this small nation. An additional benefit to the Soviets was that they were one step closer to a warm-water port in eastern Iran or western Pakistan. The natural gas is used to produce cheap anhydrous ammonia, which is, in turn, purchased by a large importer of this product in the United States.

A strange mix of business and politics, you say. It is indeed, and giant corporations and business leaders are involved. The genocide in Afghanistan and the tragic events in Poland are evidence of how U.S. money can be used against the very things we say we stand for.

Phosphates, for example, are crucial today in agriculture, and they are in limited supply worldwide. Yet our phosphate reserves in Florida are being sold to the Soviet Union as rapidly as possible. Tragically, this business-as-usual policy is draining our nation of natural resources while arming our enemies who might very well use the coming food crisis as a time to strike.

4. *Government Controls.*

President Reagan has rightly been waging a war against government controls that stultify American life. Such controls are helping to bring on the food crisis. Here's how: The world's leading food producer is the United States. According to a recent article in the *Wall Street Journal,* over half of the world's exportable food is now shipped from American farms. There is currently only a few weeks' supply of food stockpiled throughout the world. In the U.S., less than three percent of the population operate farms on less than sixteen percent of the land.

What would happen worldwide if a large number of these farms went out of business?

We will soon see, because tens of thousands of farmers are going out of business each year, many of them bankrupt. The current prices a farmer gets for many commodities are only fifty to sixty percent of the cost of producing them. Three immediate past presidents of the United States have embargoed the sale of U.S. farm products, destroying markets and effectively placing a lid on prices at well below the cost of production. The pyramiding of debts at high interest and the government control of farm prices have combined to endanger farming in America. A time of accounting must come.

Recently, I saw on television a poignant interview with two young farmers who even in grade school had nurtured the dream of farming. One of them was holding a theme he had written in the fifth grade outlining his dream. The young men, now in their thirties, had worked hard. Their crops had been outstanding, and their farmsteads were neat and clean. But now they are being forced into a liquidation sale of their expensive equipment because of the economic situation forced on them by government policy. They are receiving 1947 prices on the food and fiber that they produced for a world in need of food and clothing.

What a sad end to a healthy American dream!

When I began farming in 1947, the price of wheat was three dollars a bushel. While the cost of everything the farmer buys has increased in price from 500 to 1000 percent during these last thirty-five years, the price of wheat, and of many other commodities, such as cotton, has increased very little. No one in other segments of the economy would stand for getting absolutely nothing for his labor—and then paying for the privilege of going to work. Such is the tragedy of the American farmer.

Some of the acreage being given up by bankrupt family farming operations is being bought up by foreign investors. Some newly rich Arabs will use the land as a hedge against inflation of their petrodollars. Other acreage is going into the holdings of large, multinational corporations which seem to be able to influence price cycles to suit themselves. Many thousands of acres simply lie idle and unproductive.

The larger threat to the productive and wholesome way of life that has enabled our farms to produce food for much of the world

Table 4. Nonreal Estate Farm Debt Outstanding in the United States

Year	Commercial Banks		Production Credit Associations		Federal Intermediate Credit Banks[1]		Farmers' Home Administration		Others[2]	
Jan. 1 of	mil. $	%	mil. $	%	mil. $	%	mil. $	%	mil. $	%
1935	628	52.9	60	5.0	55	4.6	204	17.2	———	———
1940	900	26.1	153	4.4	32	0.9	418	12.1	1,500	43.5
1945	949	27.9	188	5.5	30	0.9	452	13.3	1,100	32.3
1950	2,049	29.8	387	5.6	51	0.7	347	5.1	2,320	33.8
1955	2,934	31.2	577	6.1	58	0.6	417	5.5	3,210	34.1
1960	4,819	38.0	1,361	10.7	90	0.7	398	3.1	4,860	38.3
1965	6,990	39.0	2,277	12.7	125	0.7	644	3.6	6,330	35.3
1970	10,330	43.3	4,495	18.9	218	0.9	785	3.3	5,340	22.4
1975	18,238	51.3	9,519	26.8	374	1.0	1,044	2.9	6,050	17.0
1976	20,160	50.7	10,773	27.1	350	0.9	1,772	4.5	6,350	16.0
1977	23,283	50.5	10,223	26.6	368	0.8	1,877	4.1	7,300	15.8
1978	25,709	46.3	13,508	24.3	374	0.7	3,141	5.7	8,410	14.9
1979	28,273	43.3	15,016	23.0	509	0.8	5,780	8.9	10,420	16.0
1980	31,034	41.3	18,299	24.3	665	0.9	8,982	11.9	11,720	15.6
1981	31,567	38.2	20,027	24.3	810	1.0	11,756	14.2	14,000	17.0
*1982	33,100	36.3	22,530	24.7	1,022	1.1	13,755	15.1	15,000	16.5

[1]Loans to and discounts for livestock loan companies and agricultural credit corporations.
[2]Estimates of short- and intermediate-term loans outstanding from merchants, dealers, individuals, and other miscellaneous lenders.
*(1982 figures are preliminary)

for several generations is repossession by our federal agencies. As farmers have been pressed toward bankruptcy by policies that artifically devalue our food products and force European governments into huge import duties to protect their farmers from our cheap food policies, the Farmers' Home Administration has become a lender of last resort. Now, massive numbers of farm failures within the next two years may force tens of thousands of hardworking families off the land, resulting in government ownership of millions of acres of farmland.

To make matters worse, the government is starting a pilot program to put inexperienced people with no farming background on repossessed government-owned land where they will use repossessed machinery. This smacks of the Soviet system which turned Russia—once the world's largest exporter of wheat—into the world's largest importer of grain.

How can inexperienced folks from the inner city make it in farming when experienced farmers couldn't make it on account of low prices? One thing is sure. The strength of our national economy rests on the strength of our agriculture, and that strength is being sapped by the cheap food policies of the U.S. government, high interest rates, and higher prices for the equipment, fuel, fertilizer and other essentials for agricultural production. It is a gloomy picture indeed.

5. *Economics.*

I strongly feel that there must be an increased demonstration of fairness and equity in our economic system. At this writing, the relative worth of the contribution of those who produce our food and fiber is not being recognized.

The plight of the American farmer is spreading to the general economy. One dollar in agriculture multiplies itself seven times as the price for raw materials works its way through the economy. The damage caused by a politically induced cheap food policy goes well beyond our own borders. We have not only depressed prices for farmers in other grain-exporting nations such as Canada, Argentina, and Australia, but we have undercut the livelihood of peasant farmers in underdeveloped nations as well.

The peasants of third-world countries do not have the government protection which farmers in the European Economic Community enjoy. European governments, recognizing the need for a strong agriculture, place huge import duties on American grains in order to protect their farmers from our cheap food policies. One European country recently imposed a wheat import duty of $2.50 per bushel on top of the cost of transporting the wheat across the Atlantic.

Recent parity figures announced by the U.S. government pegged the fair price of soybeans at $12.50 per bushel; wheat at $7.10 per bushel; and corn at $4.90 per bushel. Actual prices in early 1982 were half of these figures.

The farmer's percentage of national income has shrunk from eight percent in 1948 to less than one percent in 1981.

If the concept of parity (fairness) had been heeded in our policy-making, it is likely that the current debt structure which threatens our freedom could have been avoided. Parity is a natural law which cannot be artificially denied by the greed of selected sectors of an economy. The farm depression of 1927 became the general Depression after the stock market crash of 1929, and almost identical forces are at work in this decade.

Bankruptcies on and off the farm are more than doubling each year as high interest rates, often triggered by higher capital demands of large multinational corporations in various merger battles, threaten the future of small businesses, as well as the farmers who have fed our nation so well.

After a sufficient number of family farmers have been bankrupted, and economic power is further centralized in the hands of banking and oil magnates, the price of wheat on the Chicago exchange will probably start rising and keep on rising—as the price of gold did in the seventies. When this happens, the supermarket shelves will empty quickly. The panic will have begun, and food shortages will have made their appearance in the "breadbasket of the world."

The average American lives eleven hundred miles from his food supply. A major trucking strike, another increase in fuel costs, the panic of a threatened nuclear war, earthquake, drought, flooding, insect infestation, another large wave of refugees, the failure of the dollar, a chain reaction collapse of savings institutions. . . anything could trigger a breakdown in both production and distribution.

Food will be so precious a commodity then that it will overshadow the gold, silver, diamond, and oil markets. Although the price of wheat in early 1982 was $3.50 per bushel, in Revelation 6:6 the Bible speaks of a time in which a bushel of wheat will sell for a day's wage. The most negotiable currency of the world will then be grain. As the possession of oil supplies affected the world order in the 1970s, so food, or the lack of it, will influence the world order of the 1980s.

There are still those who cling to the hope that some refinement of the economic system will turn this nation's ailing economy around, and that the crash predicted by so many will never happen. But we cannot possibly have a growing federal revenue, lower taxes, budget deficits, and a permanent debt of a trillion dollars without a depression, recession, or mass inflation. It has been estimated that in 1982, forty percent of all available capital for borrowing will be required by the government. Deficits will continue and will doubtless be financed by printing more fiat money. The crash can be delayed, but can it be prevented?

For forty years we have lived on deficit spending. That means we have taken money from our children to pay our debts. That debt is now coming due, for when the food runs out, every penny will be required. Could it be possible that the day will come when wheat will be more precious than gold?

Many of us believe it will not only come, but that it is almost here.

6. *God's Judgment.*

The final reason we face a food crisis is that our nation is turning away from God in a headlong pursuit of money and pleasure. God punished Israel almost three thousand years ago for that kind of sin, and He will do the same to America unless we reverse our direction. This subject is so important that I want to dwell on it in detail in the next chapter.

2
The Forces Responsible

The possibility of God passing stern judgment upon America for our declining morality should be of great concern to every one of us. But how, you ask, does this relate to the coming food crisis? Let me explain:

Food, like all blessings, is a gift from God. When man turns his back on God, dire consequences follow. Rain is withheld; people suffer from pestilence, disease, and famine. The history of mankind is filled with such occurrences.

The concerned citizen—businessman, housewife, student, laborer, professional person—needs to understand some basic spiritual principles in order to make it through the coming crisis. If one's only purpose is to stay alive when the food disappears, one may do that for a while by robbery and murder. But if we are also interested in building a world where our children may stay alive, then it is imperative that we learn God's ways and cooperate with them.

"Our nation is in the balance between God's favor and His judgment," Pat Robertson, president of the Christian Broadcasting Network, said recently in his newsletter. "We have the choice of being a godly nation or of being humiliated and destroyed. A majority of people in the United States are not living in accordance with God's Word. The courts continue their assault on public reverence for God. Drug addiction, crime, pornography, illicit sexuality, child abuse, abortions, and divorce continue to escalate.

"Not only is our own nation ripe for catastrophic judgment by God Almighty, but the world order itself has been weighed in the balance and found wanting. The humanistic-materialistic ethic is failing. The Utopian dream of a man-made society controlling its own destiny has become a cruel joke. . . . The disintegration of Communism, the world population outstripping food supplies, a serious shift in weather cycles, the dependency on cartel-held oil supplies, a highly fragile world financial order, the high probability of war, have all served to lift the world's problems beyond the reach of man's finite abilities."

King Solomon put it this way in Proverbs 14:34: "Righteousness exalteth a nation: but sin is a reproach to any people."

And these words of the apostle Paul sound as if they had been written during the past year:

> Our time is growing short . . . Those who are enjoying life should live as though there were nothing to laugh about; those whose life is buying things should live as though they had nothing of their own; and those who have to deal with the world should not become engrossed in it. I say this because the world as we know it is passing away (I Corinthians 7:29-31, Jerusalem Bible).

Biblical prophecies foretell a coalition of Western European nations and a powerful world leader called the Antichrist. Under his rule, according to Revelation 13:16-18, world commerce will reach a degree of such sophistication that all buying and selling will be done by numbers.

If all buying and selling is to be controlled by the Antichrist, then the distribution of food will also be under his control. As we observe the international banking technology with its exchange of electronic currency units replacing the transfer of actual cash, we cannot help but accept the possibility of one man controlling all that technology. This becomes more personal when we visit a modern supermarket and observe the computerized checkout system that can charge our groceries to our bank accounts without necessitating a cash payment.

All of this has to be of special concern to the believer in Jesus Christ, yet none of it should scare the Christian, for he is commissioned to be an overcomer and to occupy this earth until Jesus returns at the sound of the last trumpet. Whereas the next few years could be a time of fear for those who do not know the Lord and have not made adequate preparation, it will be a time of great opportunity for God's children.

It is true that God promises His children ultimate safety, but because they live in a society rapidly turning against God, many will suffer some of the same earthly consequences which fall on the ungodly. However, we must remember that just as God's judgment against an immoral people has been stayed in the past, His judgment against America can be turned aside—if we act. The tide can be turned by righteous, praying people who call the nation to repentance and change.

How did we get into this situation?

It has largely come about because of this "new" religion called humanism, which has as its basic tenets that humankind is basically good, that human wisdom can solve our problems, and that we need no outside guidance or help from God. Furthermore, this religion holds that God does not exist.

Humanism was debated fully in the early days of our nation. The elitist humanism of Hamilton and others finally lost out to the Jeffersonian views based on the philosophy of John Locke and Scripture—that nations and people have a social contract with God and that there is worth in each individual, even though we all have a problem with our fallen nature.

Because Jefferson's views prevailed, we have experienced two hundred years as a nation that has been blessed abundantly. Yet as our abundance made life easy, we have gradually been taken over by a humanistic philosophy, along with the arrogance, greed, and moral decay which has historically proven humanism to be a bankrupt way of life. Any good scientist looks for empirical evidence, and yet many self-proclaimed social scientists apparently feel that more humanism is the cure for the failure of the past. They ignore the fact that there is no evidence that humanism has ever benefited any society.

A typical example of the arrogance of humanism is the insistence of many judicial and educational leaders that school children be taught that man evolved from lower creatures. While all of us in the scientific field agree that animals evolve and adapt within their species, it has never been proven that one species can evolve into another.

It has been said that the official religion of America is materialism and that her lifestyle is hedonism. If this is true, God's judgment will soon turn to wrath, and all in the nation will suffer. One of the ways we will suffer will be through dwindling food supplies.

With a food crisis on the horizon, the question which naturally comes to mind is, "Who, if anyone, can take control of the mess in which we find ourselves? If there are critical days ahead, who can help us face the future with faith rather than fear, with confidence rather than cowering?"

The Almighty is our only hope, for when God is blessing a nation, there is no limit to what can be done to correct the simple problem of food supply. Little can become much in the Lord's hands. Remember the manna God gave to His children in the wilderness? Even though they were disobedient, He chose to feed them. Remember Elijah, fed by the ravens, sustained by the widow's undiminishing supply of meal and oil? Remember Elisha's twenty loaves and sack of corn for the hundred prophets? Remember how Jesus multiplied the five loaves and two small fishes?

When God is in control, we do not need to worry about what we are going to eat or wear. All we need to do is listen to His voice, be obedient to the guidance that we receive, and be willing to leave behind some of our luxuries if we are called to do so.

I believe the days ahead will be the most challenging days ever faced by the American people—more challenging than during the Revolutionary War or the War Between the States. The best hope we have is to center our trust in God, not man, and find the center of His perfect will for us.

I lived almost forty years before I discovered this truth. In the winter months of early 1973, the burdens which I had been accumulating for four decades suddenly began to crush me. The social, economic, and political concerns of the world seemed overwhelming. In spite of my best efforts, everything about me was becoming more and more unstable. I write this with a smile because for the two previous decades I had genuinely believed that enough human effort in the right places could ultimately prevail against ignorance, disease, war, bad nutrition, corrupt power structures, and so on.

Up to that time, I had never really known fear. During the two or three times of intense danger in my life, I had reacted with calm, accelerated mental processes and quickened physical responses. So I could not understand why fear, worry, and sleeplessness had now begun to overcome me.

Underneath all this, for about four years a spiritual hunger had been gnawing at me. It led to many avenues of experimentation, but I had never felt the presence of any supernatural force over which I could not exercise control.

Tenacity was deeply ingrained in me, and it had been useful during those early, hard days of the natural-food movement, but now it was no help. The more I struggled with all of my strength against this unknown enemy, the deeper I sank into the pit of despair. I covered it well. No one knew that Frank Ford was in trouble. My pride was so strong that even my wife didn't know that I hadn't slept a full night in three months.

In 1965, and again in 1969, I had read the Bible through. It had never spoken to me, though I had tried hard to believe it. The history of Christianity had to me been the Crusades, the Inquisition, the Thirty Years' War; certainly there was nothing there that I wanted. So I just tried to be a "good" person, and like most of us, I was failing at it. But now in 1973 the situation was critical. I needed help.

Then it occurred to me that perhaps I had missed something in the Bible, so I turned to it again. In Paul's second letter to Timothy, the thought leaped out at me: "God doesn't give us a spirit of fear, but of power, and love, and a sound mind. . . ." Well, then, if God wasn't giving me this spirit of fear, perhaps some other kind of supernatural force was responsible.

I searched back through my childhood for answers. There had been a work ethic upbringing, selling produce up and down the street at age nine, working thirty-three hours a week while attending school at thirteen (in those days, stores could hire them that young). I farmed in the summer with my father, held down two jobs during my high school years, and worked my way through college. They were productive years. As the oldest of four children, I had felt a great responsiblity to my younger brothers and sister and to my mother.

My contemporaries had always seemed to like me, voting me the class president in high school. I was senior patrol leader in our Scout troop, top-ranking cadet in the largest military school in the world at that time (Texas A & M), commanding officer of a nuclear-armed rocket battery in the U.S. Army. It seemed that I was always in charge.

When I started farming, I had done it in the natural way, in harmony with nature from the beginning. I didn't know much about organic farming. I just started that way because of a gut feeling; I didn't like the idea of spraying my crops and my land with poisons. I also liked the idea of using bugs to eat bugs. I appreciated the way ladybugs gobbled up aphids. I liked the idea of building up the soil with natural fertilizers instead of harsh chemicals.

Then, because I felt my organically grown wheat was superior and ought to be specially marketed, I started Arrowhead Mills in 1960 on a weedy piece of ground at the edge of Hereford, Texas, with the money my wife and I had saved over a five-year period while living on just over $200 a month.

The three or four people who passed along the isolated road each day saw some old grain bins raised around a boot pit dug mostly through rock with an air hammer. The slope of the grain-bin floors had been shoveled out in the heat of the summer. Water had been hauled in the back of an old pickup to make the hand-mixed insulation material for the roof of the small concrete-block warehouse. An old rail car was converted into a combination office and bag-storage area. While one thirty-inch grinder turned out whole-wheat flour and cornmeal, the old pickup pulling a sixty-dollar trailer served as the trucking fleet.

In those early days, I'd drive the highway from town to town through the panhandle and south plains of Texas, stopping at grocery stores and bakeries. "This is clean, fresh, stone-ground wheat flour," I'd tell them.

The grocer would often shake his head skeptically and say,

"People won't buy this, because they think flour is supposed to be white."

Few stores were daring enough to buy the product outright, so generally I'd mark the bags with the retail price—twenty-nine cents for a two-pound package in those days—place them neatly on the shelves, and leave them on consignment.

During those seven years that our new company lost money and gave me no salary, organic farming was paying off for me. Driving a 1939 model tractor from before dawn until after dark in the summer brought financial blessings. There had been time, too, to head up a migrant ministry, to work with Scouts, disadvantaged youth, and do a number of other community jobs.

And now it was 1973, and the president of one of the most successful and fastest-growing natural-food companies in the nation was not sleeping. The man who had farmed faithfully for twenty-five years and who had a bumper wheat crop just a few months away from harvest, who had four wonderful children who were setting track records and making wonderful grades in school—this "successful" person—was in distress.

When first reading the New Testament, I had skipped the Book of Hebrews because it seemed hard to understand. But now I went back to it, and in the 12th chapter, there it was:

> "My son, do not regard lightly the discipline of the Lord, nor faint when you are reproved by Him; for those whom the Lord loves He disciplines, and He scourges every son whom He receives. It is for discipline that you endure; God deals with you as with sons, for what son is there whom his father does not discipline? (v. 5-7, New American Standard Bible).

That verse knifed through the confusion in my spirit. In total brokenness, I fell to my knees and said, "God, I don't even know how to surrender. You will have to help me do even that." At that instant, the filmstrip of self-condemnation which had plagued me for three months disappeared. I was experiencing the truth of Romans 8:1, that in Christ there is no condemnation. I realized that suddenly I was free from the law of sin and death. My debts had been paid. Jesus Christ was my personal friend.

Suddenly the intellect which had put me in the top one percent in tests given in college and in the officers' corps, and which was now so dimmed by the hell of the last weeks, was plugged into God's electricity. Through the simple, yet so difficult, act of surrendering my life to the God who had created me, I knew in my heart that Jesus Christ was who He said He was.

I knew that my God had loved me enough to send His only Son to live a pattern life, to minister to the sick and sinful, and to restore me to God through His sacrifice on that cross in Jerusalem. He had done this for *me.* I was free. Jesus had paid it all. The gift of faith was given even unto me, the one who had never believed anything he couldn't see, and didn't want anything he couldn't earn. It was free, and I had accepted the gift. I was born-again, a new person in Christ.

But the struggle was far from over. I was led to the Book of Acts. In the first chapter, Jesus told His disciples (v. 8, NASB): "You shall receive power when the Holy Spirit has come upon you; and you shall be My witnesses both in Jerusalem, and in all Judea and Samaria, and even to the remotest part of the earth." Wow! I recalled that these people had done miracles with Jesus for three-and-one-half years, yet He had told them not to leave Jerusalem until they were imbued with this power of the Holy Spirit. I figured that if Jesus' own disciples needed it, I needed it too.

The Holy Spirit who turned a group of defeated and scared disciples into the victorious army that changed the world is the same Holy Spirit who now gives me the power to share with you what I have learned from all those years of farming (more than twenty thousand hours of tractor time), from my study of the Bible, and from information gleaned from the world's experts.

In the chapters that follow are practical instructions on how you can care, share, and prepare for the greatest food crisis we've ever experienced.

3

What's Happening to Our Food?

Perhaps the saddest of American tragedies is the fact that in the richest nation in the world, millions are dying of malnutrition. Actually, the famine of real food is already here. Degenerative diseases (cancer, heart attacks, strokes, etc.) are epidemic, and are directly linked to our consumption of devitalized, additive-laden food. The statistical possibility that you will die of a degenerative disease is frightening. There is also a statistical probability that you would die sooner in this nation than in twenty-three others.

According to recent projections, one in every three Americans may be expected to contract cancer. Overall statistics that show a lengthened life span are misleading, since they do not make allowances for the tremendous decline in infant mortality. The truth is that in America the life expectancy of specific age groups—such as fifty-four year-old males—has not increased in recent years. Among the nations of the world, we rank only twenty-fourth in life expectancy, despite our technology in medicine and previous abundant food supply.

What has happened?

Most of our food is no longer true food; it is "unfood." Food processing for long shelf life has killed almost everything of nutritional value in it. The preservatives and artificial flavorings and colorings that are added are accumulating in the human system and slowly killing life in the cells of our bodies. The average American is consuming about three pounds of additives each year while some are eating as much as ten pounds a year.

The concern about preservatives is valid—food needs to decompose in digestion. Some preservatives hamper digestion and assimilation. Most preservatives are inorganic chemicals, and they are trapped in our body's system, doing to it what they are doing to the food—embalming it. Many food additives have been found to cause cancer, and the combination of so many of them can be disastrous.

Too many sweets pose an even greater problem. The refining takes out of sugar the organic acids, protein, nitrogen elements, fats, enzymes, and vitamins originally present in the plant. Hydrochloric, phosphoric, and sulphuric acids and lime are added. White sugar has been deprived of its original vitamins and minerals and has nutritive value only as a carbohydrate.

Brown sugar, corn syrup, and other refined sweeteners are also big negatives in the health equation. "Corn syrup solids," also known as commercial glucose, is one of the least desirable sweeteners and is often used because it is cheap. It is absorbed very rapidly, playing havoc with the human body. Such sweeteners add little nutritionally, actually draining the system of nutrients for metabolism. They provide the body with a "high" followed by a "low," resulting in a craving for more sugar for another "high." That's why people develop a "sweet tooth" and have increasing appetites. Excessive sugar in any form is a primary cause of hypoglycemia.

One of God's real miracles is whole grain. Unfortunately, the process of making white flour eliminates the bran and germ of the wheat kernel, along with most of the vitamins and minerals. After the bran and germ have been milled away, the flour of the powdered endosperm is bleached with chlorine dioxide gas which destroys seventy percent of the remaining vitamin E and alters the amino acid methionine to the point that it may become a toxin. Without the oil of the germ, which could become rancid, white flour can be stored for long periods. Bugs and rodents are not interested in it because they instinctively crave only what is nutritious. In short, rodents are smarter than humans—they won't touch the stuff.

Of the twenty-six nutrients lost, the "enrichment" process replaces only four—and they are synthetic.

Just as our bodies need whole grains to provide adequate fiber and good nutrition, they need unrefined whole oils which should be extracted by pressing.

Most commercial oils have been extracted by using a chemical called hexane (one of the hydrocarbons found in gasoline.) Oils which are transparent in appearance have gone through a bleaching, deodorizing, winterizing, and clarifying process which involves washing in a highly caustic solution and heating to 450° to 470°F. This process, even though it may be called "cold pressed," is hardly cold, and it removes the chlorophyll, lecithin, vitamin E, pro-vitamin A, and minerals from the oil. We should avoid coconut and palm oils, often used because they are cheap, because they are "saturated" oils which can cause hardening of the arteries.

Commercial cooking oil, margarine, peanut butter, salad dressing, and mayonnaise are usually processed further by hydrogenation. A catalyst (usually nickel) is added to the oil, which is then heated to about 365°F in a closed container while hydrogen gas is pumped through it. Again, the extremely high temperature destroys many of the nutrients.

Most Americans who drink pasteurized, homogenized milk for their good health are getting a bad bargain. The milk is heated to 161°F for at least fifteen seconds during which fifty percent of the vitamin C is destroyed, and ninty percent of the enzymes killed. Furthermore, the process alters the amino acids lysine and tyrosine, making them no longer useful to the body. Most of the essential fatty acids are also altered. Vitamins B-1, B-2, B-6, pantothenic acid, inositol, folic acid, and biotin are destroyed. Most of the calcium is oxidized and bound in a form that cannot be used in the body. All of the above healthful ingredients can be obtained in fresh certified raw milk, yogurt, kefir or acidophilus milk.

Cocoa and caffeine should be avoided. Both of them interfere with the assimilation of calcium, which is vitally important in resistance to infection and in maintaining healthy bones, teeth, nerves, and heart. Cocoa contains theobromine, a drug like caffeine, which can cause hyperactivity in children and addiction in anyone. A leading allergist considers cocoa to be the cause of many migraine headaches.

Another important key to good nutrition is the avoidance of too much salt, used in abundance in the refined foods and the fast foods of today for two reasons. First, it is cheap. Second, the taste buds of those raised on junk foods have been weakened to such an extent that it takes a heavy dose of salt to make them taste foods at all. This heavy concentration of sodium is a real killer because the body needs to maintain a sensitive balance of sodium and potassium. Many leading cardiologists agree that the elimination of salt from the diet would also eliminate up to eighty-five percent of all heart disease.

We often find such products as sodium caseinate (also used in plastics and paints) and calcium chloride (also used in antifreeze, fireproofing, and dust control) in our food supply, used as texturizers and firming agents. These and artifical colorings should be avoided completely.

It might be said, simply, that the four big killers in the modern-day diet are: 1) too much fat, especially hydrogenated and "partially hydrogenated" oils; 2) too much refined sugar; 3) too much refined flour; and 4) too much salt.

There are at least sixty known nutrients needed to sustain life. The vitamins, minerals, trace minerals, proteins, fatty acids enzymes, and complex carbohydrates can be found in natural foods. The further along the chain of processing that food has been carried, the greater the loss of vital nutrition. Boxed cereals from the large companies are often no more than super-refined confections with high percentages of sugar. They have been put through such high heat that there is little real nutrition left in them.

Our population is truly suffering from a nutrition famine, seen often in so-called junk foods and fast-food restaurants serving food that is non-nutritious. In the days to come, there will not only be a great need for food, but for real food that gives sustenance and nourishment, the kind of food that helps prevent, not promote, the degenerative diseases that are linked to the destruction of our immune systems and the weakening of our resistance to unfriendly organisms.

If there is a reader who has begun to suspect that his family's health problems are related to junk food, I urge you to seek the wisdom of God. We have a responsibility to friends and family to provide them with real food. The day might come when just a little food will need to go a long way in meeting our needs.

4
Is There Enough Food to Go Around?

We know that the abuse of good wholesome food supplies through refining, and the maldistribution on a worldwide scale have affected the well-being of most of the world's population. Waste in these and other forms has had tragic consequences.

Overpopulation itself could become a basic problem within the next few years. World population is projected to double in the next generation. Even now, one-fourth of the over four billion inhabitants of this planet are hungry, and one-eighth of them are starving. Even with the best of weather conditions, the world food supply cannot keep up with existing demand.

Ten years ago the United States was paying $3.6 billion a year to farmers to hold land out of cultivation. Now, virtually no land is being held in reserve.

Up to 1970, world fish catches had been increasing as much as five percent a year, which was then more rapid an increase than world population. However, total catches are now falling dramatically. Many commercial fishermen have had to find other occupations. Many marine biologists now feel that the global catch of table-grade fish is very close to the maximum sustainable level.

The State Department's "Global 2000" report submitted to President Carter in 1980 has a grim picture of the year 2000. According to the report, prepared in cooperation with the Council on Environmental Quality, the increasing demand for food dwarfs the harvests of available arable land. While one hectare (2.4 acres) fed 2.6 people in the early 1970s, it will have to support four by the year 2000. The report went on to emphasize that the world's tillable soil was menaced by erosion and the steady buildup of salt and alkali.

The report frankly stated that hundreds of millions of people will be hungry. For parts of the Mideast, Africa, and Asia, "The quantity of food available to the poorest groups of people will simply be insufficient to permit children to reach normal body weight and intelligence."

Famine-producing drought could shrink the world's forests dramatically, particularly in the tropics, where the poorer nations use wood for cooking and heating. The massive deforestation could raise the proportion of carbon dioxide in the air, possibly triggering climatic changes.

By one estimate, between half-a-million and two million species—largely insects and plants—could be extinct by the year 2000, mainly because of air pollution and the loss of natural habitats. Deforestation was also seen as a cause of destabilizing the world's flowing waters, silting the rivers and dams, depleting ground water levels, and aggravating cycles of flooding and drought. These would, in turn, further dwindle the world food supply. The "Global 2000" report saw a tremendous increase in real prices of all consumer goods, bringing up the question not only of whether food is available, but who will be able to afford it!

Another major problem facing us is unemployment. In depressions of the past, most of the people lived on or near the farm. Unemployment simply meant more growing of gardens. Now, with eighty percent of our population in the cities, the problem of slacking productivity and possible food shortages takes on an ominous tone.

Americans once led the world in production output per man-hour, but for the past twenty years, we have ranked *last* among the eight major industrial nations in productivity growth. Japan and Germany, with their economies completely destroyed just over thirty-five years ago, have become world leaders in productivity.

The logistics of food distribution have been greatly complicated by the skyrocketing oil prices of the seventies. The eleven hundred mile average distance that our food travels before it reaches the consumer is an expensive addition to the price of food, in time and the expense of fuel, equipment, and labor.

Packaging has become more ornate, expensive, and extensive, while contributing absolutely nothing to the nutritive qualities of the food. In fact, you can almost count on the fact that the more expensive the packaging, the less healthful the food inside it.

The millions of dollars spent in advertising low-nutrition cereals on Saturday morning television runs the cost of the average family's food budget up much more than is generally recognized. The same is true of soft drinks which have wrecked the health of millions of people. The advertising of non-food foods hurts us in a number of ways.

Inflation is another destroyer of the family food budget. Not too many years ago, a silver dollar, or a paper dollar backed up by silver, would buy eight loaves of bread. Today, a printing-press dollar will buy one loaf. The nation's leaders must make the choice between the chaos of collapse or the hyperinflation of printing more money in order to prolong things.

What does all this mean? Simply that we must return to the basics. We must become more dependent upon God and the basic gifts of soil, water, clean air, honest work, and community, that are compatible with all of Scripture.

What does it mean to get back to basics in the food realm? It means learning how to sprout whole grains in your own kitchen. It means a family garden. It means learning how to relate to whole foods instead of the refined counterfeits. It means eating less than most folks do in this country. It might mean a little place in the country and a few chickens. Most important, it means seeking God's guidance.

The coming food crisis may force many people out of the large cities. The wise man prepares ahead of time, learning now to live the simpler life.

5
Caring

Americans are developing a remarkable capacity for bad news. People have become so pessimistic and fatalistic that many shrug their shoulders at the idea of a coming food crisis.

"What can you do?" they ask. It's not really a question; it's a statement of fatalism.

Actually, a great deal can be done. First, I suggest an attitude. We have to care. We have to love people. If we do this, God can use us to help ease hurt and even save lives. What we must resist is selfishness, especially among those who claim to be believers and followers of Jesus Christ.

In the days to come, there will be tension, confusion, disappointment, shock, and sorrow. That being so, there is a need for people who will train and equip themselves now to be God's instruments in caring for His people. He has instructed us to help one another:

> As we have therefore opportunity, let us do good unto all men, especially unto them who are of the household of faith (Galatians 6:10).

Jesus taught that in the last days, we must feed the hungry.

> Then shall the King say unto them on His right hand, Come,

> ye blessed of my Father, inherit the kingdom prepared for you from the foundation of the world; For I was hungry and ye gave me food: I was thirsty, and ye gave me drink. . . . Then shall the righteous answer him, saying, Lord, when saw we thee hungry, and fed thee; or thirsty, and gave thee drink? . . . And the King shall answer and say unto them, Verily I say unto you, Inasmuch as ye have done it unto one of the least of these my brethren, ye have done it unto me (Matthew 25:34-40, Scofield Reference Bible).

It comes down to this: Are we going to be a part of the solution, or a part of the problem? Crisis always brings out the best and worst in men. I believe we will need more staying power than our own spiritual batteries can contain. The higher source comes through a vitalizing relationship with Jesus Christ. Paul describes it this way:

> I am crucified with Christ: nevertheless I live; yet not I, but Christ liveth in me: and the life which I now live in the flesh I live by the faith of the Son of God, who loved me, and gave himself for me (Galatians 2:20).

He also made this affirmation:

> I can do all things through Christ who strengthens me (Philippians 4:13, The New King James Bible).

What would God have us do? We need to pray persistently to discover His plan—it will differ for each of us—and cooperate with it. Some will be led to stay where they are, at the work they are now doing. Some will be led to concentrate on storing food, while others will be led to give themselves in helping the needy now. (If you are not being led to store food, you had better pray for the one who is. Pray that he stores enough for both of you!)

One area where the will of God for us is already very clear is that we are to make provision to care for our own family.

> But if any provide not for his own, and specially for those of his house, he hath denied the faith, and is worse than an infidel (1 Timothy 5:8).

The danger is that too many of us will undertake food storage in a selfish way. One way to avoid self-centered thinking is for us to seek opportunities to give to the poor or hungry while we conserve for ourself and family. God should get the first of everything, including money that we are thinking of spending on storage food.

To care properly for others means toughening our moral and spiritual strength. It means developing a response to crisis which includes sacrifice; it means taking adversity without complaint, suffering ridicule for our faith, and enduring physical hardship. We also need to spend more time in fasting and prayer.

We can participate in seminars and distribute printed information to help alert people. We can make ourselves less vulnerable financially by discontinuing credit cards, selling cars that are heavily mortgaged, and avoiding buying things on the installment plan.

We can store food. We can encourage our neighbors to store food. We can give storage food items as gifts on birthdays, anniversaries, and for Christmas. We can persuade churches to set up a food-storage plan in preparation for the benevolent work that will be needed when the food crisis comes.

We can help produce food. If you can't have a garden, perhaps you can have a greenhouse. If you can't have a greenhouse, perhaps you can have a sprouter. If you can move to the country and raise crops, prayerfully consider it. Perhaps a few families can buy adjoining small tracts of land and work together. We can care for the people of our nation who are going to be hurt by this crisis.

Through intercessory prayer, Rees Howells tells how a small group of regular prayer partners affected the tide of victory in World War II in Britain. God used their prayers to bless the British nation in dozens of critical moments.

Here are several more suggestions for actions that will show you care:

- Organize a prayer group to meet regularly in your home for the sole purpose of praying for the nation.
- Write letters to all your friends in other places, urging them to do the same!
- Call or telegram some official and tell him that your group is especially praying for him that day.

If you really care what happens to our beloved nation and its people, you will go into action. There is a lot that can be done. Someone who lives where you live needs to take the initiative. Is that someone you?

6
The Joseph Principle

The story of Joseph in the Bible presents a good scriptural basis for any food-storage plan. When Pharaoh's dreams (Genesis 41) could not be interpreted by the ungodly wise men of his court, Joseph, the godly man in Pharaoh's prison, was called forth. When Pharaoh challenged Joseph to interpret the dreams others had failed to interpret, Joseph said he could do so only as God gave him the ability.

Pharaoh then described a two-part dream of cattle and corn. In the first part, he had seen seven fat cows eaten up by seven lean cows. Then he had seen seven good ears of corn replaced by seven poor ears.

Joseph's interpretation was that the seven good cows and the seven good ears meant, first, a seven-year period of prosperity; the seven lean cows and the seven bad ears meant another seven years of great famine.

Joseph suggested that Pharaoh appoint a cautious and wise man to run the economy of Egypt on the following basis:

> Let him appoint officers over the land, and take up the fifth part of the land of Egypt in the seven plenteous years. And let them gather all the food of those good years that come, and lay up corn under the hand of Pharaoh, and let them keep food in the cities. And that food shall be for store to the land

> against the seven years of famine, which shall be in the land of Egypt; that the land perish not through the famine (Genesis 41:34-36).

Pharaoh so completely accepted Joseph's counsel that he appointed Joseph to head up one of the most massive food-storage programs in history. The wisdom of Joseph's advice was demonstrated seven years later when a fierce drought brought crop failure all over the Mideast. In the seven lean years that followed, there was ample corn in Egypt, and that nation prospered while others were devastated.

> And the famine was over all the face of the earth: and Joseph opened all the storehouses, and sold unto the Egyptians; and the famine became very bad in the land of Egypt. And all countries came into Egypt to Joseph for to buy corn; because that the famine was so sore in all lands (Genesis 41:56–57, Open Bible, margin).

A growing number of prophetic voices today are saying that the current abundance will soon be replaced with great lack. Difficult years lie ahead. It is not a time to be afraid, but a time to use the "Joseph principle," and prepare for the future.

A good rule-of-thumb is to set aside the equivalent of one-fifth of our food, just as Joseph had the Egyptians do. For most people today, that would mean a drastic reordering of priorities, even sacrifice, exactly as it did for the Egyptians.

Another bit of wisdom we can draw from Joseph is in the type of food to store: grain. Here was both food and seed. Seven years later, that grain became international currency, more negotiable than gold. With it, Joseph obtained money, cattle, and land. In time of famine, food was the coin of the realm.

Joseph demonstrated a daring wisdom that drastically departed from the popular thinking of his day. Today too, people who dare to be different are using the "Joseph principle" to guard against inflation, unemployment, potential disruptions of the food market, economic deterioration, and even to provide for retirement years.

God teaches that we are to gather food in time of plenty and store for times of need in other Scripture passages as well:

> Go to the ant, thou sluggard; consider her ways, and be wise: which having no guide, overseer, or ruler, provideth her meat in the summer, and gathereth her food in the harvest (Proverbs 6:6-8).

God also promises a blessing to those who sell food when people need it.

> He that withholdeth corn, the people shall curse him: but blessing shall be upon the head of him that selleth it (Proverbs 11:26).

All the fine words of sympathy that we might have for the needy will not relieve us of our responsibility. We are required to give them something more substantial.

> If a brother or sister be naked, and destitute of daily food, and one of you say unto them, Depart in peace, be ye warmed and filled; notwithstanding ye give them not those things which are needful to the body; what doth it profit? (James 2:15,16).

The world's way is to sell and buy. God's way is to give and receive. God says we reap what we sow, and there are no crop failures in God's economy. One of the surest ways to have food is to keep on giving it away. Jesus said, "Give, and it shall be given unto you . . ." (Luke 6:38).

Based on my own involvement in food storage beginning some years ago, I have a strong conviction that it is right for Christians to store food. In my enthusiasm to serve Him directly, after I was converted, I spent a lot of time away from my business, speaking on college campuses and in churches. Soon, God told me He wanted me back in my business, that He had work for me there. Then He put in my mind that I should help people set up food reserves to prepare for harder times.

I struggled with the idea, because I didn't feel that people should buy storage foods out of fear. My inner conversation with God went something like this:

"About this food-reserve idea, Lord, You know I don't believe in doing something out of fear."

"Preparation is the opposite of fear," came the answer.

"Lord," I asked, "how about people who are now going hungry? Can we store food while this is going on?"

The answer came: "Don't store food until after you have more than tithed your crop to the places I direct you. Giving must come first."

"But even then, Lord, isn't it hoarding?"

"Setting aside family-owned food reserves can provide for people in need *and* strengthen the American family farm. The whole social system will benefit."

With that, I was persuaded.

After my conversion in 1973, I wanted to be a minister. God had honored my preaching by using it to bring many to Him, including some really "hard cases." But soon after the above conversation with the Lord, I received confirmation from several mature, respected Christians in different parts of the world that I was to become a part of an end-time "Joseph ministry."

It sent me back realizing that my unique background in natural foods and natural farming had well prepared me for the coming hard times for which Christians in general were unprepared. I put Arrowhead Mills completely at the Lord's disposal.

One of our oldest customers was Shiloh Farms in Sulphur Springs, Arkansas. Shiloh Farms is a community of believers which supports itself by manufacturing and distributing natural-food products. Shiloh's business manager, Warren Clough, is also a food chemist, and we were old friends. It was Warren who first brought the subject of reserve foods to my attention. He believed that clear signs in the world, fulfilling biblical prophecy, pointed to times soon to arrive when people would need to depend upon the reserve food they had stored away.

Warren had already investigated all of the reserve-food plans on the market and found them high in price and low in quality. They were far from natural, being laced with preservatives, colorings, flavorings, and other non-food additives. In his opinion, no reserve-food programs in existence at that time answered the need for wholesome, natural nutrition. He felt, further, that our company was in an ideal position to put such a program together.

Though I was still struggling with my negative feelings about reserve foods, Warren had started something; the idea wouldn't die. The Lord kept reminding me that preparedness would help prevent fear and that preparedness did not go against the principle of trusting God. Instead preparedness means doing the possible, clearing the way for God to do the impossible.

I began to pray, seeking God's will in earnest. Early one morning, about a month after Warren Clough had introduced the subject of reserve foods to me, I had a vision. I was in Joshua Hall at Shiloh Farms at a time of national emergency, and there was a severe food shortage. A mob of about one hundred men, all armed, were massed outside, badly in need of food and ready to seize the hall by force. I had such peace in my heart that I walked outside, stood in front of them and said, "You don't need those guns. Come inside and be fed. We have food to share with you."

That vision answered my prayer. I saw with unmistakable clarity that the Christian must be like those early American settlers who always had extra food in the larder to share with passersby.

In March 1975, not long after my vision, I decided our company would go full speed ahead, creating a program of whole foods in their natural state, uncontaminated by spray residues or food additives. Our foods would contain the calories, vitamins, minerals and enzymes necessary to support good health, and we would offer them for sale at a reasonable price. And so it was that The Simpler Life Reserve Foods came into being.

Something that had happened about a year before had stuck in my mind. I didn't fully understand it at the time, but as we got underway with the reserve-foods program, it began to make more sense. In 1974, while visiting the Berkeley campus of the University of California, I was invited to a local prayer group by a Christian I had just met. As the meeting was about to end, the elder in charge came over to me, laid his hands on my head, and began to prophesy. He had never met me. He didn't know my name, where I was from, nor the nature of my business. Yet he gave the following message:

> You will be used to feed millions of people. I will take the food

> and divide it among My people. You must be obedient and not compassionate. You must be obedient, listen for guidance, and not be compassionate.

I was sure that the prophecy was genuine, but at the same time, I was a bit puzzled. I had always been taught to *be* compassionate. After prayer, the Lord showed me that *He* would be compassionate. All He wanted from me was obedience to prepare a food ministry and wait for the Lord to awaken others to the need.

7
What Foods Should Be Stored—and How?

Why do you spend your money for that which is not bread?. . . Hearken diligently to me, and eat what is good (Isaiah 55:2, Revised Standard Version).

Isaiah's question about spending money on non-food was used as a figure of speech, but I wonder how he would react if he saw the way people were buying non-food nowadays! He might not know much about vitamins and minerals, but he could not fail to see the folly of buying something that would hardly nourish an animal, much less a family.

If we're going to use hard-earned wages to buy storage food, let's be sure we're getting something that will keep us healthy. There are four tests you might use to determine if the food is worth buying:

One: Will it support life? If insects won't eat it, neither should you.

Two: Does it contain additives? If it has additives, don't store it.

Three: Is it unprocessed? Most processing, especially that involving heat, radically alters the food. In the case of grains (except for hulled rice, oats, and barley) and beans, a good test of whether you're getting whole food is whether it will germinate and sprout if you plant it.

Four: Does it help supply essential vitamins and minerals? Foods should be chosen that will complement one another, providing vitamins, essential fatty acids, carbohydrates, minerals, and proteins.

In general, foods that are low in oil content—such as wheat, beans, potato flakes—will last much longer than foods with more oil—such as almonds, peanuts, sesame seeds, and pecans. Both freeze-dried and dehydrated items can be expected to keep their good taste and nutrition for from five to ten years, as long as the cans are kept intact. The shelf life of these foods is just a few months once the cans have been opened and then reclosed with plastic lids.

According to several sources, whole wheat and powdered milk will keep indefinitely when stored properly. More conservative estimates say that we can expect whole grains, beans, flakes, and certain other reserve foods to be still in good condition fifteen years or longer after purchase.

You can store sea salt, honey, teas, and—for a shorter time—cooking oil, peanut butter, natural fruit juices, and other items according to anticipated cooking and beverage needs. Add freeze-dried foods, dehydrated fruits, cheese powder, and such snack items as banana flakes to give greater variety to your menus.

Be sure to deal with food suppliers who market wholesome, natural, organically produced foods and stand 100 percent behind their products. For long shelf-life, make sure their high quality foods are put into enameled cans packed in nitrogen atmosphere rather than having nitrogen inserted with a probe.

In planning your food-storage program, remember that proteins are the body's basic nutrients for growth and energy. Without adequate protein, your body cannot utilize the vitamins and minerals necessary for good health. There are twenty-two different amino acids in a whole protein, eight of which cannot be produced by the body. Only a few foods in the American diet contain complete protein. Eggs contain all twenty-two amino acids, but some are lost in cooking. For a whole, balanced protein, store grains like oats and wheat, and legumes like soybeans and lentils which are rich in protein and relatively inexpensive. Eating a combination of legumes and grains in the same meal will take care of your body's needs for protein.

When buying milk powder for storage, be sure it is labeled non-instant, nonfat. Instant powdered milk has been subjected to a heat process that makes it more like plastic than a wholesome food product.

Consult Appendix I for a list of foods to store, and the bibliography in the back of this book for further basic information to guide you in planning your food-reserve program.

You will want to store foods that can be easily prepared in a simpler life-style than the one to which many Americans have become accustomed. Keep in mind what your neighbors or friends would need if you shared some of the unprepared foods with them.

In times of emergency, electric appliances such as mixers, blenders, and food choppers will probably be inoperable unless you have your own electrical generator. Hand-operated grain mills, food choppers, and blenders might be purchased now before panic drives up the prices on these items. A handmill will make it possible for you to turn your grain into cereal or flour when there is no electricity.

We may have to do without refrigerators. Stoves may have to be woodburning or kerosene. A good cast-iron pot with a cover could prove useful. There may be no running water. In view of these facts, some storage foods need to be precooked, freeze-dried foods.

Of course you need to store foods that you can afford to buy. Food prices are a comparative bargain now. Consider whatever you buy a kind of food insurance, a good hedge against the coming inflation.

You would have to have an above-average income to be able to store a year's supply of nothing but freeze-dried foods, which cost three to five times as much as nitrogen-packed grains and beans. The prices of freeze-dried foods vary greatly among the different companies; unfortunately, quality also varies. Steer clear of foods containing sulfites, BHA, BHT, coloring agents, refined ingredients, and TVP (textured vegetable protein, a byproduct of chemical oil extraction).

Actually, you may be able to afford to buy more storage foods than you think. If you gave up one meal a day to reduce your present grocery costs, for instance, you could use the money saved to buy food reserves. Most Americans would be healthier on just two meals a day anyway.

Pyramid selling schemes should be avoided in purchasing foods, as the extra profit margins make basic foods too expensive.

Getting together with others in your church fellowhip can often enable you to get lower prices on a food reserve. Available discounts may run from twenty percent upward when groups combine their orders.

Perhaps there are current expenses in other areas of your family's budget that can be reduced to enable you to begin a food-storage program. What about turning down the thermostat and the hot water heater, unscrewing some light bulbs, storing electric hair dryers that are running up the utility bill, cutting down on long-distance calls, consolidating some of your errands into town to save gas, or carpooling to work or school, and eating out less often?

Some people have used the cash value of their life-insurance policies, sold the second car in the family, their boat, or a silverware set to get started reserving food. Others are having garage sales to finance food storage programs.

How important will food storage be to your family? It may well be a matter of life or death.

Along with food reserves and other necessary household goods, there is a need to make provision for pure water. Disruptions in your water supply caused by war or other disasters can take a terrible toll if you are not prepared to provide your own safe drinking water. Without potable water, one can live only about three days, while the body can go for long periods without food. For that reason, some device to remove harmful bacteria and chemicals from your water should be part of your planning. There are many products on the market making astounding claims, but some of them are not backed up by proper testing. A good water purifier will contain granular activated carbon and silver. Read the labels and shop intelligently and carefully.

Of utmost importance is the proper packaging of storage goods. Wet-packed canned goods of the kind you normally find in the supermarket will keep for a few months. They take up more space and are less economical than dried foods because of their water weight, but they are generally readily available. If you are buying storage food, however, it is important to invest in something that will retain its value over a longer period of time. Nitrogen-packed grains and beans will last many years—much longer than a crisis is likely to exist—and they will take up only one-fourth to one-third the space of wet cans containing the same amount of food.

Hardly anything in a pasteboard box is worth storing, and the box only runs up the cost of foods for current use. Basic foods should be purchased in clear plastic bags so you can see the quality of the food you are getting.

For freeze-dried and dehydrated food reserves, the packing is already done. However, you may want to store some air-dried peas or corn or anything similar that you have bought in bulk. For do-it-yourself food packing, you can recycle small containers from coffee, shortening, honey, and other foods. For storing small quantities of grains and legumes, containers with reusable plastic lids are suitable, as are other glass jars with screw-on lids. Gallon jars made of glass or plastic are especially good, and can also be used as they are emptied for homemade sprouting gardens.

Ask neighborhood restaurants, school cafeterias, and such places to save larger containers for you. Number-ten cans, gallon size, will hold from five to seven pounds of grains or legumes. Never use any container that has held petroleum, chemicals, or building materials. They may have absorbed harmful residues, and they may be made from vinyl chloride or other harmful materials which can be absorbed by your food. For safety, recycle only containers that have already been used for food products.

If you can't round up enough free containers, you can buy new ones. A four-gallon can will hold twenty-five pounds of wheat; a five-gallon can will hold about thirty-three pounds. Half-gallon glass canning jars hold about three pounds of wheat per jar and have future canning value.

New metal trash cans, properly sealed, can be useful in storing dry grains and legumes. You can also purchase five-gallon plastic containers with resealable lids at most natural-food wholesalers.

Where should reserve foods be stored? The answer should take into consideration temperature, humidity, accessibility, safety, and ease of concealment should the need arise.

For optimal storage life and freshness of your foods, keep them in the coolest and driest part of your house, basement, or garage. The ideal temperature for the storage of dehydrated and freeze-dried foods is between forty and seventy degrees Fahrenheit. Attics, top shelves of closets, lofts, and such places are undesirable locations because of their extremes of temperature. Crawl space under the house in most areas would be ruled out for the same reason. Foods will do better in the part of the house where you live. Low humidity is important, so avoid placing containers directly on a concrete floor or against a concrete wall.

8
Methods of Food Preservation

To the pioneers who settled this country, food storage was an important part of the life-style. Perhaps you even had grandparents who remember when everyone got most of their food right out of the field and stored their own supply for the winter. Some of the old-timers can tell you about shelling corn and grinding it for cornmeal and then storing it in barrels.

Our grandparents salted meat down in tightly closed boxes or smoke-cured it. Fruit was dried on wooden trays covered by cheesecloth and left in the sun. Peas were dried in the hull and then threshed and winnowed, to be placed later in airtight cans for storage. Vegetables were canned in glass jars. Sweet potatoes were piled on the dirt floor of a barn and wrapped snugly with warm straw and old quilts to prevent freezing. Fruit was canned in its own juice in glass jars. These were the staples. With a few chickens for fresh eggs and cows or goats for fresh milk, a family was assured of food for the winter. But that was another life-style in another time. What can be done today?

The food preservation industry has been revolutionized in the last ten years by the development of the freeze-drying process by which virtually all moisture is removed from foods while preserving garden-fresh flavor and vital nutritional content. Foods are flash-frozen at minus 50°F., then placed in a vacuum chamber where radiant heat turns the frozen water content directly into a vapor, which is literally vacuumed away. When all the water is removed, the foods are packed in a special nitrogen-filled atmosphere in heavy-duty sealed cans enameled on the inside and double lacquered on the outside for rustproofing. Freeze-dried foods keep their taste and nutritional value for several years.

The quality of freeze dried foods is usually such that many people are unable to tell them from fresh, regularly prepared foods, and they are easily and quickly prepared by simply adding water. For warm food, simply add hot water. Most of these foods can be ready to eat in five minutes.

The method of putting up your food reserves will depend, in part, on the amount you want to store. If you have less than one hundred pounds, you can use the freezer treatment. Simply fill your containers, seal them tightly, and leave them in your freezer for a month or more. According to the United States Department of Agriculture, common insects are killed in from two to thirty days in a freezer that registers from 0° to −20°F. You can place filled containers in your freezer for a longer or shorter period, depending on the temperature of your freezer and the size of the container.

Another way of treating small quantities of food for long-term storage is with low oven heat. Set the temperature of the oven at 140°F. Spread the grain in pans to a depth of not more than three-fourths of an inch, and heat at 140° for thirty minutes. It may be necessary to prop the oven door slightly ajar to prevent scorching around the edges. Wheat treated in this way will not germinate. According to the Department of Agriculture, insects are destroyed in two hours at 120° or thirty minutes at 140°. Place the heated grain or legumes in clean, dry containers and tightly close or seal them. For larger quantities of grain, use the "dry ice" treatment, about eight ounces of dry ice per hundred pounds of grain, or about half an ounce per gallon. Do not handle the dry ice with bare hands because it will burn. Watch children closely to see that they do not try to play with it or eat it. Use extreme care when using glass containers, because if you miscalculate and seal them too soon, the glass can explode from the pressure. Here is a good procedure to follow:

Pour one or two inches of grain in the bottom of the container. Then place the prescribed amount of dry ice on the layer of grain and fill the container as completely as possible with more grain. Place the lid loosely on the container. Do not even begin to tighten it. After the piece of dry ice has been entirely vaporized, tighten the lid securely.

As the dry ice evaporates, gas forms that is heavier than air forcing all the air from the container, and leaving an oxygen-free atmosphere in which insects cannot live and in which their eggs will not hatch. The gas will remain in the container for some time, even with the lid fitting loosely, so you do not have to be in a hurry about tightening it. Better to be too slow than too soon. Keep the container sizes and the amount of dry ice placed in each one uniform so you can use one container for test purposes.

To test your dry ice processing, use a nail or an ice pick to make a hole in the lid, then follow the above procedure for filling the container. Wait for about thirty minutes, then tighten the lid and place a drop of water over the hole. If it makes a bubble, carbon dioxide gas is still being released. If not, seal the hole with tape and tighten the rest of the containers after waiting the same length of time on each of them. With glass containers, wait several hours.

Sealing the cans and jars that have been treated in the above methods is not difficult. A tight-fitting screw-on lid sufficiently seals the container. Most snap-on lids for plastic buckets will adequately seal, as well as the lids of metal trash cans. For extra security, you may use rope caulking, and as a further precaution against lids being accidentally pried off, seal with heavy tape.

Don't forget that you can preserve many foods by home canning for only pennies a jar, and that you can sun-dry or dehydrate with a homemade dehydrator for almost no cost at all.

Finally, let me encourage you to rely mostly on foods which are dried in the fields by the sun and breezes. These foods are more nutritious and less expensive, and there is something of a continuing miracle in the way God has made provision for us by giving us "every grain bearing seed," according to Genesis 1:29. These "grains bearing seed" have almost unlimited storage life and are the product of God's special love and care for His creation. In over a quarter of a century of active farming, I have never ceased to be amazed at how beautifully balanced nature is—when we allow it to be.

9
Moving to a Simpler Life-Style

One thing is certain: The food crisis is going to force us back to a simpler life-style. We are going to have to adjust to much more than a scarcity of food.

We are moving into a time that will demand more self-efficiency in dozens of areas—from baking our own bread to building our own homestead. Whatever cars we have will necessitate a basic acquaintance with automotive mechanics. We may have to administer more than simple first aid in case of injuries or sickness. We may need to know how to do basic plumbing repairs, carpentry, or masonry.

If the flow of consumer goods and public utilities service is interrupted, items like wood heaters and kerosene lamps may be needed. Bicycles will also be in great demand. Emergency equipment like fire extinguishers, flashlights, and battery-operated radios may soon become necessities.

Women may need to learn again to make their own dresses, can their own food at home, and "make do" without the dozens of aerosol household products that now seem so indispensable. In fact, many items that we take for granted every day might simply not be available.

In view of these possibilities, we must attempt to store more than food. We are going to need some equipment and supplies with which to begin our simpler life. Some recommended items:

- transistor radio
- battery
- warm clothing
- tool kit
- personal toilet articles
- matches
- candles or oil lamps
- garden seed
- fire extinguisher
- port-a-potty
- first-aid kit
- sprouting kit
- grain mill
- flashlight
- batteries
- towels, blankets
- rope, twine, wire
- soap
- condiments
- stored water
- water purification tablets
- large plastic bags
- toilet paper
- garden tools
- wood stove
- food dehydrator

You'll need to put a few drops of bleach in each gallon of stored water. Large plastic bags will serve several uses—among them, disposable raingear and fallout clothing! They also would be helpful in moisture-proofing boxes of stored foods should you have to haul them or store them temporarily outside.

Not only would it be a good idea to stock up on a few items we must have for weathering a crisis, we may also need something for use as barter. Barter is the age-old currency that takes over when money fails. In an emergency, it may be the only way to buy or sell.

Bartering is also a fine way to help one another because it is a form of sharing. We ought to get into the habit now. Start a list of people and what they have for trade (don't forget skills), in your community or your church. Bartering is also one way you still might be able to purchase food when the crunch comes. Don't rely too much on this, however. If the food isn't there, you can't get it even with barter.

You would be wise to start storing hardware items such as small tools, garden implements, nails, fence wire, hand-operated grain mills, bicycles, camping equipment, car batteries, tires, and tire repair kits. These will make excellent barter items since they will probably be in great demand. Also, household items such as razor blades, first-aid kits, kerosene lamps, spring-operated clocks, and shampoo will make good barter material.

Right now you may be thinking, "Thanks! Now I *know* I can't afford to get involved in this. I don't have the kind of money I'd need to buy all that." Here's a basic rule that will be worth remembering: Never pay the retail price. Take advantage of flea markets, garage sales, pawnshops, the Goodwill and Salvation Army thrift stores, and wholesale catalogs. Usable goods, whether new or used, will work better than money in the near future. Just think of it as a new way of banking, and start collecting what you can when you can.

Although our primary use of stored food outside the family circle might be to provide for the needy, and in most cases without expecting payment in return, we may need to do some bartering with food items.

Preparation for a move to the simpler life includes planning for alternate energy sources. Since an energy crisis may be linked to the economic and food crisis, there is a distinct possibility that our electric stoves may not be functional when the need arises to prepare meals from stored foods.

Wood is a renewable energy source that is enjoying a resurgence in popularity. If you have your own woodlot or can obtain free firewood, you can save money now by using wood heat. Newspapers often carry ads for free firewood to anyone who will cut and clear unwanted trees. Sawmills have scrap heaps where firewood can be picked up free. Highway construction frequently requires the destruction of trees that can be cut up and hauled away just for the asking.

The best woods for fuel are hardwoods like oak or hickory, and firewood is more efficient when it has been allowed to "cure" for a while after it is cut. Most cities have firewood companies that are doing a thriving business since wood is the most common source of alternate energy.

If you are installing a wood heater as an alternate source of heat, it would be wise to check the efficiency ratings of the various types and brands of heaters on the market. It would also be wise to choose a wood-burning stove with a flat top that could be used for emergency cooking.

Another energy source is methane gas. Methane is produced biologically, from manure. If you live in a rural area, you might want to obtain information on a bio-gas plant that you can build with scrap materials. A simple arrangement that can provide gas for cooking and hot water for a family of four requires the presence of only three or four cows or pigs.

The "digester" is a 200- or 300-gallon tank that keeps the slurry on low heat while the methane gas formed by bacteria flows through a pipe to the area to be heated. A more complicated arrangement can produce methane that fuels an electric generator.

In some areas, wind energy can provide electricity to power household appliances. A prevailing wind pattern of at least twelve miles per hour is necessary for most units. Setting up the tower and the turbine itself are the most expensive part of this system—usually costing several thousand dollars. However, there can be some pretty rewarding dividends. Some homeowners who have set up their own wind-generated plants not only have adequate power for their own needs but are now selling surplus electricity back to the local power company!

Swift-running streams can make it possible to use hydroelectric power as an alternate energy source at much less expense than a wind generator. Many people living in mountain areas already have private electric plants operating from streams on their property.

Portable emergency gasoline-powered generators are also available. These are self-contained and require only gasoline and a connecting power cord to be operational. Whichever energy source you choose, be sure to have some means of cooking food when the power goes off.

Another part of the changeover to the simpler life will be in learning different eating habits. Unless we are already accustomed to eating whole, unprocessed foods, this will mean a big change. The best approach is gradual. Reduce spices and flavorings to allow your taste to adjust to natural flavors. Work on getting a balanced diet.

Soybeans are about the cheapest form of high protein available, and they have a fairly neutral taste. That means you can combine them with other foods, and they will take on the flavor of whatever you are cooking with them. Keeping a supply of flaked or ground soybeans in your kitchen will make the job of putting wholesome meals together a lot easier. They're good in spreads, patties, breads, casseroles, soups, and in sprout salads.

Good whole-grain bread will be a mainstay in the new diet. If you buy it, do so at a good natural food store. Since most of the grain in commercial whole-grain loaves is heavily fumigated, the label is printed "no preservatives." One hundred percent whole-grain flour should be the first ingredient listed on the label. Start experimenting with making your own bread from whole-wheat.

Since beans and peas can be bought and stored in bulk, begin introducing them to your family in a variety of recipes. *The Simpler Life Cookbook* should be very useful. For the best in natural foods cookbooks, try the *Deaf Smith Country Cookbook.* Both are available in natural-food stores. *The Living Cookbook* by Yvonne Turnbull is also excellent.

Make frequent use of sprouting grains, beans, and other seeds. A small homemade sprouter can be made with a glass jar with a perforated lid or covering. (See Appendix.) More efficient ones are sold at health-food stores for a few dollars. For a family that may not have regular access to fresh garden greens, sprouting is an excellent means of having fresh vitamin-packed salads year round.

Yogurt is easy to make at home, and easier to assimilate than milk. For help in getting started, recipes are provided in the Appendix to this book.

The move to the simpler life will put us more in touch with the outdoors. The back-to-the-land movement in the United States is partly an expression of the need to be basically more self-sufficient. It is a good urge; if we can't take care of ourselves, we will be in trouble. But it is also a movement caused by a craving to be closer to nature—to walk in the woods, swim in a clear stream, climb a mountain.

Whether you move your residence to the country or not, it would be nice to get some experience backpacking, trail-riding, hiking, and camping out. These activities will also help you keep your body physically fit.

The Master Engineer has designed us in such a way that our bodies improve with activity. We are not mere machines, which wear out with use, but we are living beings that must have exercise to keep from wearing out. When you eat, you stack up calories (energy units) that are meant to be burned up by the exercise of your muscles. If you eat, you should work. On the other hand, according to II Thessalonians 3:10, ". . . if any would not work, neither should he eat."

If we abuse our bodies with overwork, friends are quick to say, "You should not work so hard; it's not good for you!" But if we neglect our bodies by not working, no one dares to say, "You should get out and work more!"

The sedentary occupation of some overweight persons complicates their weight problem, but that does not excuse us from the need to teach and tone the muscles of our body. Use the stairs rather than the elevator. Walk to lunch. Jog in the morning.

Research has shown that fifteen minutes of exercise in which the heartbeat rate is increased to between 140 and 160 beats per minute (depending upon your size) will result in a slightly increased metabolism rate during the entire twenty-four-hour period that follows. This is an encouragement to us who have thought we were burning up fat only when we were performing the actual exercise. A workout early every morning has benefits all day, not the least of which is increased efficiency of digestion and the glandular system.

More activity out-of-doors will help us get or keep our bodies trimmed and toned. It will also sharpen our appreciation and concern for our environment. Over the past few years, there has been an increasing awareness of the need to conserve our natural resources and protect the quality of life that God gave us on this planet. Slowly, we are beginning to realize the need to reduce our demands upon the ecosystem and to live more closely in harmony with God's laws.

Our forestlands, lakes, rivers, and streams are all interlinked in a complex system that affects our water supply and the agriculture that brings food to our table. In any move back to a simpler lifestyle, we should do what we can to promote soil conservation, reforestation projects and the cleaning up of streams by stopping the dumping of chemicals into our rivers.

We can study local wildlife and the habitat that it requires, and help ensure that the habitat is preserved. If we are privileged to own a piece of ground in the country, we can commit ourselves to farm it in such a way that we enhance its fertility, rather than mine it out. One good thing the food crisis is certain to bring about in this nation is a greater concern for the soil.

The simpler life is a better life. If it takes a crisis to move us into it, then the crisis won't be totally bad; it will have been useful in the larger purposes of God for our lives.

10
The Community of Sharing

"No man is an island," and all of us need other human beings. In Psalm 68:6, the Psalmist writes, "God setteth the solitary in families." We need to be needed, and we need to have a sense of belonging. A crisis will always bring neighbors together in a unique bond that may not exist at other times.

There is, however, a community of sharing that can have that bond at all times. It is composed of true believers in the Lord Jesus Christ, a group which follows His teachings about a life of giving. There are thousands of such groups operating throughout the world, with more being formed every day.

The world tells us to reach out and grab for ourselves all that we can. Jesus says, "Give, and it shall be given unto you." Jesus is a Giver, and He wants us to be like Him. It is so important to Him that we learn to give, that He rewards us richly every time we do.

It is a basic truth that whatever we sow we reap, but according to the law of the harvest, we always reap more than we sow. We always get back more than we give.

In the same way, love demands demonstration in giving. If we don't have a giving nature, we don't have a loving nature. The Body of Christ (His followers as described in I Corinthians 12) is so interdependent that when one part has a need, all the others feel it and rush to respond.

In this type of Christ-centered group or community, if a widow needs food, the more fortunate members respond in love and provide for her. If a member of the Body has a hospital bill he can't pay, the other members of the Body come to his aid. It is as simple as that. Galatians 6:10 admonishes us, "Do good unto all men, especially unto them who are of the household of faith."

If we do the possible, God will do the impossible. We are to help one another in the ways that we can. Then God takes over the rest of the project, and miracles happen!

The more we give, the more God wants to encourage and prosper us. And when we have a personal crisis, other trained "givers" are there to be sensitive to our needs. According to Acts 2:44, 45; 4:34, 35, that's the way it was in the first-century church. It worked so well, "neither was there any among them that lacked." I believe that Christians must again become that kind of closely knit, mutually supportive people.

We dare not neglect to learn these principles. Can you imagine the shameful spectacle of greedy and panic-stricken Christians gulping down hoarded food while others were starving? We must learn to share now in a time of available supply, so we can do it in an atmosphere of panic.

If we did not have friends who had learned to share, who would help us if we were robbed or our reserves confiscated? It is in our own best interest to become as generous as we possibly can right now, and to pray that we can infect many around us with the same spirit.

Another basic principle strongly impressed upon me: We should never look to our stored food for security. If we begin to depend on it instead of on God, we will make it necessary for Him to remove the food from us. If we look only to God, as His obedient children, and see the food simply as His to use, He will see that we are always supplied.

> And God is able to make all grace abound toward you; that ye, always having all sufficiency in all things, may abound to every good work (II Corinthians 9:8).

In days of crisis, the community of sharing can help sustain us in more than physical needs. When we learn to really care about one another, we are concerned about every aspect of need. We can turn to the community of sharing when we feel the need for direction. The other members of the Body can help us pray for direction from God.

In individual Christians and in meetings of Christians, the gifts of the Holy Spirit called words of knowledge, words of wisdom, and prophecy will often be given in direct answer to prayer.

I received such direction back in 1975 when I was confused about the direction God wanted me to go. At that time, I happened to visit a church in California. During the service, several elders stood up and gave short but strong prophecies about me. They spoke of my having a ministry like that of Joseph and Noah in the Bible. In one of the prophecies, the phrase "watchman on the wall" stood out. It brought to mind a Scripture:

> I have set watchmen upon thy walls, O Jerusalem, which shall never hold their peace day nor night: ye that make mention of the Lord, keep not silence (Isaiah 62:6).

A watchman had a position on the wall where he could see what others could not. That being so, he was required to give warning and could not keep silent. Later the same year in Arkansas, the same phrase occurred in another prophecy that was given to me.

> Thou shalt have an understanding of the times, as did the children of Issachar, and receive a prophet's ministry to become a *watchman on the wall*.

The next year, the phrase was repeated in still another prophecy given in a prayer group in Texas: "You are a *watchman on the wall*." Those identical messages, given in three different states, were a tremendous help to me in settling down to the work God called me to do. The prophecy is being fulfilled in my life now and through this book.

I have many friends who have in the same way found direction at a critical time, when the Holy Spirit dropped the word they needed into the mind of some other member of the Body, and that other member faithfully delivered it. Usually it was confirmed and repeated two or three times in different places.

Being in a community of sharing can keep us from becoming

discouraged when things go wrong. No one can be productive when he is down. The community of sharing can help us when we hurt. In the healing atmosphere of loving acceptance and support, with encouragement from those who have already gone through similar trials, we are helped back on our feet. In the process, we have learned how good it feels to have someone care. Then we can turn around and start helping someone else. That's the way it ought to be.

At this point, some reader may say, "I need to be a part of a community like that. How do I get in?"

Well, I have good news for that person. The answer is nothing complicated like having to apply for membership. The ticket of admission is a way of life.

People who belong to such communities accept the reality of Jesus and believe that He took their place on a cross to keep them from having to serve out an eternal sentence in hell for transgressing the laws of God. When these people have expressed remorse in their hearts for being sinners and have asked Jesus to forgive them, they are given new life. The new Presence they feel is the Spirit of the Lord Jesus Christ. That's what we call being born again.

Being reborn spiritually makes one a part of the great spiritual family of God, which is part of the community of sharing that we have been talking about.

There are three aspects to a healthy spiritual state: being born of the Spirit (John 3:5), filled with the Spirit (Ephesians 5:18), and walking in the Spirit (Galatians 5:16). No one can be filled with the Spirit who has not been first born of the Spirit, and no one can walk in the Spirit without first being filled with the Spirit. The ability to walk daily in cooperation with the Spirit of Christ is a sign of spiritual maturity.

I share these things in the hope that you may find the inner peace and power I have found in Jesus Christ. Once you have found it, find your place in the community of sharing.

It is my prayer and hope that the message of this book has come into your hands at just the right time to be truly helpful to your understanding of why the food shortage will occur and what you can do about it. Welcome to the ranks of a growing army of those who are caring, sharing, and preparing!

SECTION II
Simple Foods and How to Prepare Them

1
God's Food—God's Way

We as a nation have not used the spiritual resources which have been given so abundantly by our Creator.

Our bodies, which the Bible calls the "temple of the Holy Spirit," have been abused with the rest of our environment because we as individuals and as a nation have failed to repent, to accept His grace, and to live in the power of His Spirit. Earthly resources are running out, but the resources available to us from our Lord are limitless. He has made over 7,000 recorded promises to us. One of these is found in 2 Chronicles 7:14. It is an oft-quoted verse closing with a fantastic statement: "If my people . . . turn from their wicked ways . . . *I will heal their land.*"

Many families are becoming interested in organic gardening these days. By using companion plantings such as marigolds, onions, or garlic to reduce the insect problem, and by using compost on their soil so they do not have to use chemical fertilizers, they are finding they can enjoy quality produce from their own yard—and at the same time help nourish the fragil web of life that covers our earth.

I have made an extensive study of the basic foods necessary to sustain life. Those that cannot be grown can still be readily purchased and easily stored. Here is a full list of the available basic foods for good nutrition.

WHEAT, HARD RED WINTER. There is an age-old saying that

"it takes nine months to make wheat or man." Because the hard red winter wheat grown in Deaf Smith County takes nine months from seed to crop, the roots have time to go deep into the mineral-rich subsoil of this area, producing a grain noted for its protein, vitamin, and mineral content. Wheat is the basic food of Western civilization.

RYE. This excellent grain is often used for variety in breads, and while it does not have the gluten content of wheat and therefore does not rise like wheat bread, it does have an excellent taste. Rye is grown in many areas of the Great Plains.

YELLOW CORN. Many people now own their own hand grinders and grind their own grits, meal, or flour fresh each day. Yellow corn is high in Vitamin A and makes a delicious meal. Corn and beans have been a staple food of much of the Western Hemisphere.

BROWN RICE, LONG AND MEDIUM GRAIN. Most commercially available rice has not only been grown with practices potentially very damaging to the environment, but has also been polished with a process of removing much of its B-vitamin and mineral content. The brown rice grown with sound agricultural practices is nutritious and is a delight to prepare in soups, casseroles, and many other dishes.

BROWN RICE, SHORT GRAIN. This rice is more glutenous than long and medium grain and has a delicious flavor. Rice has long been the staple food of the Far East as well as a much-appreciated delicacy throughout the world. It is easily digestible as well as nutritious and lends itself to many uses, including baby food.

TRITICALE. This is a true grain, which has been developed from hard red winter wheat, durum, wheat, and rye. It is slightly higher in protein than wheat and has a better amino acid balance than many grains. As it is low in gluten, it should be mixed with wheat flour for bread, but it makes an excellent pancake mix, cereal, whole or sprouted grain for good nutrition.

WHEAT, SOFT. Soft wheat is used for pastry flour, and its smooth texture can be used in many dishes, from the batter for tempura vegatables to the finest cakes and piecrusts. It is a different grain from hard red winter wheat in that it has a softer, fatter kernel.

BUCKWHEAT GROATS. Buckwheat is grown in the Pennsylvania-New York area. The best grains, after hulling, are sold as groats. White buckwheat groats are unroasted, and brown groats are roasted. Buckwheat is high in B vitamins and is highly regarded for use in many dishes, including pancakes.

MILLET, HULLED. Known as the "poor man's rice" in much of the world, millet is a very nutritious, low-gluten grain. It is grown in North Dakota and Colorado and makes a good ingredient in soups and stews as well as a breakfast cereal. It is a staple food in Africa, China, and Japan.

BARLEY, PEARLED. Barley has a tight-fitting hull which is removed by the process called "pearling." It is grown in the Red River Valley of North Dakota and Minnesota and is pearled as lightly as possible to remove the fibrous hull yet leave maximum nuitrition. Barley is delicious in soups and casseroles.

SOYBEANS. Grown in China for 3,000 years, soybeans are now grown so extensively in the U.S. that three-quarters of the world production is in this country. Soybeans should be cooked or sprouted to destroy the urease and anti-trypsin enzymes which interfere with digestion.

MUNG BEANS. These small beans are used almost exclusively for sprouting, have a high germination rate, and are very easy to sprout. The delicious, crisp sprouts can be used in cooked vegetable dishes (especially Chinese dishes), soup, and salads.

BLACK-EYED PEAS. Like other legumes, black-eyed peas are excellent soil builders. They take the nitrogen from the air and convert it into soil nitrogen through nodules on the roots. Black-eyed peas originated in Asia and are a favorite dish of many people, both by themselves and as an ingredient in soups or stews.

LIMA BEANS. First grown in tropical America, lima beans, like other beans and peas, are a good source of protein. Better utilization of protein is obtained by eating beans and grain dishes in the same meal, balancing the essential amino acids.

LENTILS. The most famous lentil-growing areas of the United States are western Idaho and the neighboring area in Washington

State. Lentils are an excellent source of protein and have a variety of uses, including sprouting. Cooked, they are often used in soups and casseroles.

GARBANZOS (ALSO KNOWN AS CHICK-PEAS). This legume is a traditional food of the Middle East. Like most beans, they are a good source of protein, minerals, and vitamins. They are raised in Mexico and in California and are often used in soups, pates, vegetable dishes, and salads.

SESAME SEEDS. Rich in protein, unsaturated oils, and calcium, the sesame seed is grown in Mexico and Central America. It makes a good complement to many beans and breads and is the source of high-quality cooking oil favored by many cooks.

SUNFLOWER SEEDS, HULLED. One of the nice things about sunflower seeds is that they are not only very high in protein, unsaturated oils, and minerals, but they are also delicious as a snack or as an addition to an almost unlimited number of dishes. They can be included in breads, cookies, main dishes, salads, and homemade granola.

ALFALFA SEEDS. While a package of alfalfa seeds might seem expensive at first glance, people who have seen how much delicious sprout salad can come from a spoonful of seeds are turning more and more to this easy-to-sprout seed for healthful salads.

PEANUTS. Just eighty miles southwest of Deaf Smith County, the famous Valencia peanuts are grown in an average of 350 days of sunshine a year. Peanuts are a legume rich in protein, unsaturated oils, and, of course, flavor.

ALMONDS. Grown in California by a dedicated group of growers, almonds have been called the "king of nuts." They are delicious as well as a good source of protein. Slivered almonds are often used in granola as well as in many other recipes.

CRACKED WHEAT CEREAL. This cereal makes a hearty meal that costs only a few cents. Prepared by cracking grains of hard red winter wheat into six or eight easy-to-cook particles, cracked wheat cereal is a favorite with children. Whole-grain nutrition is a good way to start off the day.

BULGUR-SOY GRITS. Bulgur-soy grits are a combination of two basic foods—bulgur, which is wheat which has been partially cooked under pressure, and the versatile soybean, which furnishes additional protein. The nutty flavor of these two together makes a delicious breakfast or side dish.

WHEAT FLAKES. These flakes, made with hard red winter wheat, are produced with a unique flaking system which utilizes a short-duration dry-radiant heat which partially cooks the grains while at the same time preserving vitamins and minerals. They are easily stored, kept cool and dry, and quickly prepared in many ways. They are also a good snack just as they come from the package.

RYE FLAKES. As rye is used less than wheat in breads and other baked goods, the use of these nutritious rye flakes in soups, main dishes and granola-type cereals is a good way to take advantage of the many fine qualities of rye. The dry-radiant flaking method has been tested and found to retain the nutrition of the whole grain.

TRITICALE FLAKES. The outstanding nutrition available in this new grain, along with the taste of these dry-radiant-produced flakes, makes this a favorite for inexpensive, yet delicious, eating. These flakes can be used in a vast variety of soups, casseroles, and other nutritious dishes which are easy to prepare.

SOYBEAN FLAKES. Prepared by the dry-radiant process, these meaty flakes cook more quickly than whole soybeans, making them a favorite for many types of main dishes. The high protein of soybeans makes this a real food bargain at a fraction of the cost of other protein foods. Inventive and budget-conscious cooks have made soy flakes a staple in their kitchens.

MAPLE NUT GRANOLA. The careful roasting of a tasty blend of wheat, rye and oat flakes, sliced filberts and almonds, sunflower and sesame seeds, and just the right amount of corn germ oil, all in a generous portion of pure maple syrup from New England, with raisins added lavishly after baking, has made this ready-to-eat cereal a favorite for snackers, packers, and all kinds of other folks. Good on ice cream or yogurt.

STONEGROUND WHOLE WHEAT FLOUR. The careful stone grinding of the best quality wheat available has made the art of home baking with whole wheat flour come alive again. The delightful aroma of bread dough rising in the kitchen or baking in the oven gives the whole family an incentive to be close by when the bread comes out of the oven.

RYE FLOUR. An interesting and tasty change of pace can be accomplished in the oven with a Swedish or Italian rye bread. Rye flour can be mixed with whole wheat or with cornmeal to make some interesting pancakes, too. The unique flavor of rye has made this flour a respected member of the flour-bin group.

YELLOW CORNMEAL. It even makes fish taste better, say the fishermen. For others, stoneground cornmeal is at its best in a tasty cornbread steaming under some butter and perhaps topped off with honey.

TRITICALE FLOUR. Fast becoming a favorite with many people, mixed with whole wheat flour in breads or for use in pancakes and waffles, triticale flour provides the good nutrition of whole-grains along with a unique taste provided by the wheat and rye from which it was developed.

SOY FLOUR. Soy flour has been lightly roasted by dry-radiant heat to improve its digestibility and flavor before being ground into fine flour. Soy flour is a good protein ingredient for almost any bread or pancake and can be used in many other dishes as well.

RAW WHEAT GERM. Wheat germ is the embryo of the wheat kernel and is responsible for carrying the spark of life to the next generation of wheat. It is a very nutritious addition to bread doughs, cereals, casseroles, salads, or main dishes.

WHOLE WHEAT PASTA. Noodles, spaghetti, and other pasta made from special types of whole-wheat flour, and variations made with rice and soy flours, make up a line of easily stored, nutritious food which can be prepared at a moment's notice. All the pasta recipes can be prepared with whole grain pasta as the base.

TAMARI SOY SAUCE. This soy sauce, made with soybeans, wheat, and sea salt, naturally fermented and aged in wood for two years, is a delicious addition to vegetables, casseroles, soups, grain dishes, or salads. It is a live food, made without artificial chemicals and preservatives common to commercial soy sauce. Tamari soy sauce is used often in the recipes in this book because it adds nutrition as well as taste.

SEA SALT. Unrefined sea salt, which has been solar evaporated and briefly kiln dried, contains essential trace minerals as well as sodium chloride. No synthetic iodine or other chemicals have been added. Excessive salt of any kind should be avoided.

PEANUT BUTTER. Peanut butter is a rich source of protein, and when spread on whole wheat bread provides a well-balanced protein. It is easily stored and makes a nutritious as well as delicious addition to almost any meal. Most commercial peanut butter is hydrogenated (a process considered harmful to health by many medical experts) and contains additives. Old-fashioned peanut butter is just pure sun-dried Valencia peanuts, famous for their taste.

PRESSED COOKING OILS. Unrefined pressed cooking oils are oils which smell and taste like the grain or seed from which they were pressed. Refined oils, some labeled with the misnomer "cold pressed," are usually light, clear, odorless products which have had much of the nutrient value stripped away. Using good quality unrefined expeller pressed oils is as important as using good quality whole-grain flours. Most unrefined oils should not be heated over 350°F. For deep frying up to 400°F., safflower oil should be used. A blend of several oils, such as safflower, soy, and peanut oils, provides a better balance of unsaturated fatty acids than does any single oil.

DRIED FRUITS. Dried apples, apricots, peaches, and pears—along with dates, prunes, and raisins—are excellent for satisfying a sweet tooth in a way that is beneficial to health. They make an excellent after-school snack and are almost essential for camping and hiking trips. They can be eaten as is or soaked and added to breads, pancakes, and dessert dishes or fruit salad.

SEA VEGETABLES. Many essential minerals and vitamins are found in abundance in food from the sea. Sea vegetables such as dulse, nori, and wakame can be a nice addition to many dishes, especially soups. They come packaged dry in case you don't live close to the ocean. Soak the dried sea vegetable for about fifteen minutes before using in your recipe.

APPLE CIDER VINEGAR. There is a lot of interest in the healthful properties of apple cider vinegar, but it should be made the old-fashioned way from apples which have been grown with natural methods.

BEVERAGES. For those who like a hot drink to pick them up, a variety of herb teas include a taste or two for every palate. Camomile, peppermint, rosehips, spearmint, comfrey, lemongrass, and various blended flavors, have beneficial effects in the body. A squeeze of lemon juice makes them even better. Several of these herb teas make good iced tea.

You probably have some favorite fruit juices. Pure fruit juice is not only good for you, when taken in moderation, but the taste of a glass full of ice-cold tangy grape or apple juice on a hot day surely hits the spot. As people turn more to plenty of water, taken between meals, pure juices, and herb teas, many of the health problems caused by some of the highly advertised drinks in this country will be reduced.

SWEETENERS. We know that the human body is designed to handle the natural foods which were the common diet until just a few years ago. There is ample medical evidence that the refined sugars and flours which make up a huge portion of the food intake of this nation today are having a disastrous effect in individual lives and in society as a whole, as medical costs skyrocket. Within the concept of whole natural foods, raw unrefined honey, maple syrup, and molasses are popular, though they should be used with moderation.

WHEAT BRAN. Wheat bran can be added to any soup, casserole, bread, salad, or dessert, either in the main recipe or perhaps as a garnish, to ensure enough fiber in the diet.

2
Edible Wild Foods

Did you know that the little dry pods which form on wild rose bushes after the delicate roses have dried up and the petals have fallen to the ground can be picked and stored all winter? Did you know that they could be munched on, ground into powder, or brewed into tea? Most important, did you know that these little rosehips are perhaps fifty times as rich in vitamin C as some citrus juices? Historians estimated that scurvy, caused by vitamin C deficiency, has killed many millions of people. Most of these deaths occurred in winter or on board ships at sea. These deaths could have easily been prevented had people known the nutritional value of rosehips.

Many people caught in the wilderness without adequate food supplies have been saved from starvation because they knew that the sappy layer under the bark of an aspen or cottonwood tree was edible, or that the fruit of a prickly pear or yucca could make a tasty snack. In the hardwood forests, hickory nuts, native pecans, acorns, black walnuts, and other nutritious nuts abound. In the coniferous forests, pinon pine nuts can keep you going. Wild cherries, plums, and apples, along with blackberries, raspberries, and gooseberries, provide a good selection in some parts of the country. The leaves of wild strawberries and grapes can be eaten as well as the fruit.

In the desert, a barrel cactus has quenched the thirst of more than one dry traveler, and chia seeds are famed for their nutritional value. In the swamps, wild rice is a delicacy which brings high prices when it can be found in specialty markets. From the ocean, Irish moss, dulse, laver, and kelp are rich in many minerals essential for good health. In middle America, the sunflower provides good nibbling. Jerusalem artichokes and cattails are widespread and very nutritional.

Mint tea has refreshed many thirsty folks, and a few leaves of mint are good in almost any fruit juice or tea. Watercress is good either raw or cooked, and a surprising number of "weeds" can be counted on to sustain life or provide a fresh salad in a pinch. These include dandelion, lamb's-quarters, pokeweed, clover, mustard, shepherd's purse, chickweed, and mountain sorrel. Some are best steeped in hot water a few minutes, but they can be eaten in many ways and can be improved by a bit of horseradish or a few wild onions.

There are precautions that should be taken if one is going to munch on wild foods which might not be familiar to the muncher. Hard shiny berries should be avoided unless one knows about them, for instance, and wild onions should smell like onions, as there are plants similar in appearance which might be toxic. There are many good books on the subject, one of the most authoritative being *Edible Wild Plants,* by Oliver Perry Medsger (see Recommended Books).

Most Americans have been overprotected from the real world by a very fragile world of technology. God placed many good things at our disposal on this earth, and you never know in advance when He might require you to know something about them. John the Baptist did very well in the wilderness. You can be prepared to do it, too.

3
Sprouting

In order to obtain the maximum benefit of the grains, beans, and seeds called living foods, one can sprout them to increase vitamin content and activate enzymes. It is important that grains, beans, or seeds be cleaned extremely well before sprouting.

A wide-mouthed glass jar is suitable for sprouting. Drop a tablespoon of seed covered with water in a pint jar for six to twelve hours. Place a piece of cheesecloth or nylon netting over the mouth of the jar and secure it with a rubber band. After twelve hours drain off the soaking water, rinse with warm water, drain thoroughly, leave jar tilted with mouth down. Room temperature and indirect light works fine. Rinse and drain each morning and evening—perhaps more often if humidity is extremely low, or extremely high.

After the sprouts have reached the desired length, place in direct sunlight for a few hours to develop chlorophyll, then rinse with cold water, drain thoroughly and store tightly covered in the refrigerator. Eat them as soon as possible, preferably within two or three days. For a fresh and continuous supply, start a new batch every day or so. Be sure seeds are obtained from a reputable food company and have not been treated. Do not eat potato sprouts as they are toxic. Sprouting is a great activity for children. Our daughter, Cindy, won a school science award in fifth grade with a sprouting project. Sprouts are fun to grow and taste good, too. Alfalfa sprouts are great in salads and sandwiches. The heavier bean sprouts are really tasty when sauteed with onions and other vegetables.

Best wishes on your sprouting adventure!

4
Yogurt and Sourdough Recipes

The best use of milk, whether it be from the cow or goat, is the making of yogurt, except, of course, for feeding infants who cannot receive enough milk at their mothers' breast. Yogurt is easily digestible, and the beneficial bacteria in it multiply at body temperature, producing B vitamins and overcoming many mild stomach upsets. Yogurt is especially helpful in restoring intestinal flora after the use of antibiotics. It can be eaten plain or with fruit or honey. It can be used in place of sour cream dressing for many dishes. Granola and yogurt go well together.

How to make your own yogurt:
1 quart fresh milk
3 tablespoons or more of yogurt

Be sure all utensils are scalded. Heat milk to scalding. Cool to lukewarm. Blend in a little non-instant milk powder if thicker yogurt is desired (optional). Stir yogurt into the lukewarm milk. Pour the mixture into scalded glasses or jars. Place jars in large pan of warm water. Maintain at about 110 degrees for 3 to 5 hours until the new yogurt thickens. Refrigerate until used.

Sourdough Bread

Sourdough bread takes perhaps twice as long to rise as yeast dough, but it is actually easier to make as well as very tasty, depending on the quality of the starter. Sourdough also makes great rolls and pancakes. Many campers have taken some starter along on camping trips and had a good breakfast working for them through the night.

How to make your own sourdough starter:
½ package dry yeast
1 cup lukewarm water
1 cup whole wheat pastry flour

Dissolve yeast in water and let set for 15 minutes. Slowly add flour and mix well. Place in scalded jar. Let set at room temperature for several days, until fermented and bubbly. Stir well and refrigerate. Starter is now ready to feed for making bread. It is best to feed starter every week or so. If the starter ferments and liquid separates, just stir to blend, feed it, and use it.

Feeding your sourdough starter:

1 cup of starter
2 cups warm water
2 cups whole wheat pastry flour

Always use an earthenware bowl to let starter rise in. Do not use metal. Beat all ingredients together with a fork. Remove fork. Let set overnight. Return 1 cup of starter to refrigerator. This leaves about 3 cups of starter to use—enough for some pancakes and 2 loaves of bread (see sourdough bread and pancake recipes).

Take Sourdough to Camp

1 cup sourdough starter
Whole-wheat flour—take plenty

Knead enough flour into the starter to make a soft dough. Sprinkle the dough with more flour. Carry the dough in a tightly closed plastic bag containing a cup of dry flour. Sourdough starter must be mixed with flour because in liquid form it can only be stored in glass or ceramic containers, which are too heavy and inconvenient for camping. The starter will keep a week or more and can be kept fresh by feeding and using it once a week or more.

To use this starter: Place the soft dough starter in a bowl. Add 2 to 4 cups of warm water and 2 to 4 cups of flour. Set the bowl in a warm place 3 to 4 hours or overnight (near a warm fire in the winter). Now the starter is ready to use in any sourdough recipe.

Sourdough on a Stick

Sourdough bread dough—
Or thick sourdough pancake batter with enough flour added to make a soft dough
Or dough made from plain pancake batter using less liquid than called for
Green sticks, peeled
Hot coals from campfire

Prepare the dough. Roll pieces of dough into long skinny sausages about ¼ inch thick. Wet the green stick and heat it over the coals. Wrap the dough around the warmed stick. Hold the dough over the coals until baked, turning to prevent burning. Dip the dough in a mixture of melted honey, butter and cinnamon before baking, if desired.

Sourdough Pancakes

Serves 4

Feed your starter an extra amount the night before.

2 cups sourdough starter the consistency of a heavy
½ cup warm water
2 Tbsp raw honey
2 tsp sea salt
2 tsp low-sodium baking powder
2 eggs
4 Tbsp unrefined oil

Dissolve honey and salt in warm water. Add baking powder, then stir mixture into starter. Beat in eggs and oil with a spoon until the mixture is smooth. Bake on a medium-hot, oiled griddle.

Primitive Bread

Sourdough bread dough
Plenty of green leaves (maple, aspen, bay, magnolia, birch, etc.)
Hot coals

Allow the bread dough to rise once. Pat the dough into ½-inch thick cake. Place the dough on several thicknesses of leaves. Scrape coals and ashes to one side. Place the leaves and dough on the hot fire base. Cover the dough well with more green leaves, then with ashes followed by a layer of hot coals. Let bread bake under the ashes and coals 10 to 15 minutes. Test after 10 minutes by poking the bread with a long thin twig. If it comes out clean, the bread is done.

Sourdough Batter Bread
Yield: 1 loaf

2 cups sourdough starter
1 Tbsp raw honey
1 Tbsp unrefined oil
1 tsp sea salt
2 Tbsp soy flour
3 to 4 cups whole wheat flour

Feed starter the night before. Stir all ingredients together except flours. Mix dry ingredients and slowly beat into batter. Beat for 5 minutes. Pour batter into a greased 9 x 5 inch loaf pan. Let rise until batter is just over rim of pan. Bake 50 minutes at 350°F.

Sourdough Bread
Yield: 2 loaves

1 cup sourdough starter
2 cups warm water
1 to 4 Tbsp raw honey
1 Tbsp sea salt
½ cup unrefined oil
6 cups whole wheat flour

Feed your starter the night before. Let it set overnight. Return some starter to refrigerator. Use large ceramic bowl or crock. Beat together all the ingredients except flour, then add flour. Mix well, by hand if necessary. Cover dough and let rise 3 to 4 hours or until doubled. In summer, it may take less time. Shape the dough into 2 loaves, kneading a little as you shape them. Place in oiled loaf pans. Let rise 1 to 2 hours until doubled. The bread will rise a little more while cooking. Bake 30 to 40 minutes at 350°F. Turn out on a rack to cool.

5
General Recipes

A brief note about the following recipes:

The ingredients have been combined for maximum protein effectiveness, overall nutrition, ease of preparation, and appealing flavor. These recipes are designed for camp cooks, chefs who want to get into natural foods, mothers who want to prepare delicious food in just a few minutes, and anybody else who would like to get away from the plastic world of highly processed, denatured foods. Use your own skills to modify, substitute and improve as you go along in the exciting world of natural-foods cookery.

(P.S. Over the past four decades, I have enjoyed approximately 400 days and nights of camping. During this time, I have built stone ovens in the mountains and used easily carried reflector ovens on the canoe trails. Even the more complicated recipes which follow can be used in the wilds. However, it would be best to save them for the kitchen, and use the simple ones in the woods.)

How to saute vegetables (or grains):

Heat a heavy pan to moderate heat. Add a good quality unrefined oil (see "Pressed Cooking Oils" in Section II) to coat bottom of pan. Add chopped vegetables (or grains) immediately and stir lightly with a wooden spoon or spatula. After vegetables have heated through, reduce heat and let vegetables cook in their own juices until tender. Start with tougher ones first, adding later those which are more tender. In dishes which require subsequent simmering, saute in a pot large and deep enough to add water, and save cleaning an extra utensil. In nearly any soup, as well as in grain and vegetable dishes, sauteing in a good unrefined oil prior to simmering locks in taste and tenderness as well as adding the extra nutrition of the oil.

How to add water:

"Doesn't that sound ridiculous?" I said to my wife, Margie, who is coauthor of the bestselling *Deaf Smith Country Cookbook.* "Anybody knows to add enough water to keep something from burning."

She answered, "Good cooks want to know how much water to start with."

Since she knows how much I don't know about cooking, I am going to compromise, do what she says, and list for you the approximate amounts of water and lengths of time required to cook the various whole grains, flakes, cereals and beans:

Per 1 cup of dry	Add water	Cover and simmer about
Barley	2½ cups	60 minutes
Millet	3½ cups	30 minutes
Oat flakes	3 cups	10 to 30 minutes
Rice, short grain	2½ cups	50 minutes
medium grain	2 cups	50 minutes
long grain	1½ cups	45 minutes
flaked	2 cups	10 to 20 minutes
Rye	2½ cups	60 minutes
flaked	2 cups	10 to 20 minutes
Wheat	2½ cups	60 minutes
flaked	2 cups	10 to 20 minutes
cracked	3 cups	10 to 20 minutes
Chick-peas (soak overnight)	4 cups	2-3 hours
Lentils	3 cups	1 hour
Pinto beans	3 cups	2-3 hours
Soybeans (soak overnight)	4 cups	3-4 hours
Soybeans, flaked	2 cups	1-2 hours
Split peas	3 cups	45 minutes

Note: The above approximations, which will vary with altitude and several other factors, allow for cooking up all the free water. For soups, more water and longer simmering times are required. The more people who drop in unexpectedly, the more water you may use. Excess water should never be poured off, as it contains vitamins and minerals from the food cooked in it. If there is an excess, save it for soup stock.

Natural Foods Are Fun

The following recipes have been adapted from the *Simpler Life Cookbook.* Some have been modified to provide easy and nutritious uses of storage foods such as freeze-dried scrambled eggs, freeze-dried fruits and vegetables, and the dairy and meat products that can be stored along with grains, beans, and seeds. In other cases, the recipes include garden vegetables and fresh eggs, current foods you might be able to provide for yourself.

BREAKFAST ADVENTURES

Scrambled Eggs with Sprouts

Serves 4

2 cups freeze-dried scrambled eggs
1 cup onion dices
1 Tbsp unrefined oil
1 cup mung bean sprouts
½ tsp sea salt or cayenne pepper

Stir water into freeze-dried scrambled eggs until proper consistency is obtained. Add seasonings and oil. Stir in sprouts and heat in skillet, stirring gently until hot. This recipe is good with freeze-dried beef or turkey dices for additional flavor.

Quick Whole Wheat Pancakes

Serves 4

2 cups whole wheat flour
½ tsp sea salt
2 tsp low-sodium baking powder
2 cups milk
2 fresh eggs, beaten well
2 Tbsp unrefined oil

Mix dry ingredients. Mix liquid ingredients and add to dry mix. Stir, adding a little water if necessary for proper consistency. Bake on oiled griddle or heavy skillet using medium heat.

Camper's Breakfast

Brown rice, whole grain wheat, or other whole grains;
Pinch of sea salt; boiling water; thermos

Use a thermos or heavy saucepan with a heavy, tight-fitting lid. Fill thermos 1/3 full of whole grain. Fill the remainder of the thermos with boiling water. Seal immediately and let set overnight.

Bulgur Breakfast

Serves 8

1½ cups of bulgur-soy grits
½ tsp sea salt
4 cups boiling water
½ cup wheat germ
¼ cup sesame seeds
¼ cup chopped almonds

Stir grits and salt into water. Reduce heat, cover, and simmer 10 to 15 minutes. Add other ingredients and simmer a few more minutes. Add apple dices, date sugar, or raw honey.

Peanut Butter Flapjacks

Serves 4 to 6

2 cups whole wheat pastry flour
½ tsp sea salt
1 fresh egg, slightly beaten
2 cups milk (or water)
2 Tbsp date sugar
2 Tbsp unrefined oil
1 Tbsp yeast dissolved in ¼ cup warm water
¼ cup old-fashioned peanut butter

Mix flour and salt. Mix egg, milk, date sugar, oil, and dissolved yeast. Gradually stir liquids into peanut butter until smooth. Add to the flour mixture and stir just until moistened. Let the batter set for 15 minutes. Batter will be thin. Bake on hot griddle until browned on both sides.

Wheat and Sesame Cereal

Serves 4

1 Tbsp unrefined oil
1 cup cracked-wheat cereal
¼ cup sesame seeds
¼ cup corn germ
¼ cup date sugar
3 cups water
Optional: ½ tsp sea salt, dried fruits, chopped nuts

Heat heavy pan. Add oil, cracked wheat, sesame seeds, and corn germ. Saute until light brown. Add water, salt, fruits, and date sugars. Cover and steam about 25 minutes. Add nuts to taste. Serve with milk and honey.

Quick Soy Pancakes

Serves 8

1½ cups soy flour
½ cup cornmeal
1 cup whole wheat flour
2 Tbsp low-sodium baking powder
½ tsp sea salt
2 cups milk (or water)
5 fresh eggs
1 cup blueberries

Mix dry ingredients. Beat liquid ingredients together and stir flour mixture into liquid. Refrigerate overnight. Add about 2 cups warm water in the morning, stir in blueberries, and bake on medium-hot, oiled griddle.

SALADS TO SAVOR

Taboolie

Serves 8

2 cups cracked-wheat cereal
  or bulgur-soy grits
1 cup chopped parsley
½ cup chopped onions (green,
  if available)
2 tomatoes, chopped
¼ cup lemon juice
½ cup unrefined safflower
  or olive oil
salt/pepper to taste

Pour 1 cup boiling water over the cracked wheat. Let sit 15 to 30 minutes. Add other ingredients and toss together. Chill before serving. Don't miss this easy-to-prepare taste treat.

Tossed Salads

Wash greens and vegetables quickly in cold water and dry with paper towel. Chill if not used immediately. Tear greens and chop other crisp vegetables. Toss immediately in enough unrefined oil to seal edges. Only then add chopped wet vegetables such as tomatoes, and choice of dressing. This procedure will prevent vitamin loss and keep the salad crisp.

Use all kinds of colorful crisp vegetables to add nutrients and beauty to salads. Rely heavily on sprouts. Growing your own guarantees you cheap, fresh, unsprayed greens all year round. They add variety and taste to your diet, as well as vitamins, minerals, and important enzymes. Raw vegetables are an essential part of a good diet.

Lentil Salad
Serves 8

2 cups lentils, rinsed
4 cups boiling water
½ cup unrefined safflower or olive oil
minced garlic
sea salt or tamari soy sauce
½ cup apple cider vinegar
paprika or cayenne

Add lentils to water and simmer, covered, about 30 minutes or until tender. Drain and cool. Mix oil and vinegar and stir lightly into lentils. Season with garlic, salt, and paprika. Refrigerate and serve chilled with tomatoes and lettuce or alfalfa sprouts.

Marinated Mushrooms
Serves 4

1½ cups sliced mushrooms
¼ tsp sea salt
¼ tsp cayenne
3 Tbsp lemon juice
⅓ cup unrefined olive oil

Rinse and slice mushrooms (or reconstitute if they are freeze-dried). Mix seasonings, lemon juice, and olive oil. Stir gently into mushrooms and let stand at room temperature for 2 to 3 hours.

Soybean Salad
Serves 6

3 cups cooked soy flakes
½ cup green pepper dices
½ cup onion dices
½ cup tomato granules

Here is a good way to use foods from your reserve food supply. Reconstitute freeze-dried vegetables, add to cooked soy flakes, season to taste, and chill.

Sprout and Grow Salad
Serves 4

2 cups alfalfa sprouts
1 cup tomato granules
½ cup green pepper dices
½ cup onion dices

Reconstitute granules and dices, toss lightly into sprouts with oil and vinegar dressing. Serve immediately or chill.

Sprout and Vegetable Salad
Serves 4 to 6

2 cups mung bean sprouts
1 cup mushroom slices
½ cup green pepper dices

Reconstitute slices and dices, toss with mung sprouts with a little unrefined oil and yogurt dressing.

DRESSING YOUR SALADS

Green Goddess Yogurt Dressing

Makes 1 cup

½ cup yogurt (fresh or reconstituted powder)
1 ripe avocado
1 tsp sea salt
1 Tbsp lime juice
1 tsp chili powder
1 small clove garlic, chopped

Optional: chopped chives, parsley or mint. Mash and blend all ingredients together. Chill.

Buttermilk Salad Dressing

Makes 3 cups

2 cups buttermilk
¾ cup cottage cheese
¼ cup cider vinegar
1 Tbsp Italian herbs

Beat together and store in covered jar. (If buttermilk is not available, use a combination of milk and yogurt powder from the reserve food units. Freeze-dried cottage cheese is also available in the long-term storage cans.)
NOTE: Your supply of unrefined cooking oils should be obtained locally at a natural health-food store, kept cool in a dark place, and used and replaced on a rotating basis every few months. You can purchase corn germ oil, olive oil, peanut oil, safflower oil, sesame oil, soy oil (rich in lecithin) and sunflower oil (and in some cases even wheat germ oil) *unrefined* (not "cold-pressed," which *is* refined).

STURDY SOUPS

Soups can be made from almost any good food, and are truly the best way to eat in many cases where fancy baking and other cooking methods dependent on technology might be unavailable. Grains and beans, cooked in the same soup, seasoned properly, can provide total protein in many delicious variations.

Barley Soup

Serves 8

2 Tbsp unrefined oil
1⅓ cups pearled barley
½ cup soy flakes
½ cup onion dices
½ cup celery dices
½ cup mushroom slices
¼ cup beef dices
5 cups water

Saute barley in the oil; add water and bring to a boil; add soy flakes and simmer for 45 minutes or so. Add the freeze-dried dices and slices, and season to taste.

Celery Soup

Serves 4

1 cup celery dices
½ cup onion dices
½ cup carrot dices
3 Tbsp parsley flakes
2 Tbsp soy flour made into a paste with ¼ cup water
4 cups water

Mix all ingredients and simmer 5 minutes.

Spanish Bean Soup

Serves 8

2 cups dry pinto beans
1 cup wheat, triticale, or rye flakes
2 tsp cayenne
2 Tbsp unrefined oil
1 cup onion dices
½ cup green pepper dices
8 cups water

Add pintos to boiling water, cover and simmer 2 or 3 hours, until almost tender. Add wheat flakes and simmer another 20 minutes. Add remaining ingredients and more water if necessary. Turn off heat and stir a minute or so.

Millet Stew

Serves 8

6 cups boiling water
2 cups hulled millet
1 cup potato flakes
1 cup carrot dices
½ cup onion dices

Add millet to boiling water. Cover and simmer for 20 minutes, or until tender. Add remaining ingredients, season to taste, stir well, and serve.

Split Pea Soup
Serves 4-6

1 Tbsp unrefined sesame oil
1 onion, diced
2 stalks celery, chopped
4 cups water
1 cup dried split peas
½ tsp sea salt
Pinch of marjoram
1 bay leaf
1 Tbsp tamari soy sauce

Heat a soup pot, and saute in oil, onion, and celery. Add water and bring to a boil. Add peas and simmer 45 minutes. Add salt, marjoram and bay leaf and continue cooking 45 minutes or longer. Add tamari. Remove bay leaf before serving.

Soybean Chili
Serves 6

1 cup soybeans or other beans
1 cup whole grain wheat
6 cups water
1 Tbsp chili powder
¼ tsp cayenne
1 clove minced garlic
1 cup onion dices
1 cup green pepper dices
1 cup tomato granules
2 Tbsp unrefined oil
Tamari soy sauce

Soak beans and wheat overnight in plenty of water. Next day cook them in a covered pot 3 to 4 hours. Add remaining ingredients. Simmer until beans and wheat are tender. Season with soy sauce or salt. Garnish with onions.

Sprout Soup
Serves 4-6

4 cups soup stock
2 cups fresh soybean sprouts
Tamari to taste
2 tsp salt
½ cup freeze-dried scrambled eggs

Garnish: scallion greens, sliced on long diagonal. Bring stock to a boil. Add sprouts and salt. Simmer for 10 minutes. Remove from heat. Add eggs, stir and heat through. Add seasoning and garnish.

Whole Grain or Flake Soup

Serves 6

3 Tbsp unrefined sesame or safflower oil
1 large onion, thinly sliced
3 stalks celery, finely chopped
3 cups cooked whole grain (rice, buckwheat or millet) OR 1 cup uncooked whole grain flakes
4 cups broth, vegetable stock or water
1 Tbsp chopped chives
1 Tbsp chopped parsley
1 tsp sea salt

Heat large saucepan. Saute in oil, onions, and celery. Add cooked grains, or uncooked grain flakes and saute. Add broth and seasonings, cover and simmer 30 minutes, or until slightly thickened. Correct seasoning to taste.

Vegetable Soy Sesame Soup

½ cup unrefined oil
1 cup carrots, diced
½ cup celery, chopped
Green pepper, chopped
Tomato, chopped
½ cup sesame seeds
½ cup whole wheat flour
2 cups cooked soy flakes
6 cups water
¼ cup Tamari

Saute vegetables and seeds in oil. Stir in flour. Add flakes and water, simmer till tender. Add tamari. Season to taste.

MAIN DISHES

(NOTE: Complete, balanced protein can be obtained by eating meals containing both a grain and a bean dish, or a dish containing both grains and beans.)

Sesame Green Beans

Serves 4

4 cups green beans or Italian green beans or any crisp vegetable which is eaten cooked
1 Tbsp unrefined sesame oil
1 Tbsp sesame seeds
Sesame salt or sea salt to taste

Cook beans in minimum amount of water. They should be slightly crisp when done. Drain any excess liquid and save for sauces. Heat a skillet on medium heat. Add the oil and sesame seeds and stir until seeds begin to sizzle. Add the beans and stir gently until heated through. Season to taste.

Limas and Vegetables
Serves 6

4 cups boiling water
2 cups dry lima beans
½ tsp pepper
1 tsp sea salt
2 Tbsp unrefined oil
1 cup carrot dices
1 cup onion dices

Slowly add beans to boiling water. Reduce heat, cover, and simmer 1½ hours or until almost tender, adding more water if necessary. Add other ingredients and continue cooking until all is tender.

Corn Mush
Serves 8

1½ cups ground corn grits
½ cup soy flour
1 tsp sea salt
½ tsp sage
6 cups cold water

Mix meal, flour, sage and salt. Stir in water slowly. Bring to a simmer. Cover and cook about 45 minutes. Include a few chopped jalapeno peppers and tamari if you like.

Grain Tuna Casserole
Serves 6

2 cups water or stock
2 cups cooked millet
2 cups cooked rice or other cooked grain
1 cup tuna dices
¼ cup unrefined oil
2 cups vegetable dices
½ cup cheese powder
Season to taste

Mix all ingredients. Place in oiled 2-quart casserole and bake for 20 minutes at 400°F. This is a good recipe for using leftovers.

Sprout Burgers

Yield: 1 dozen ½-inch thick patties

2 cups soybean sprouts and wheat sprouts mixed
½ cup sunflower seeds
½ cup sesame seeds
2 cups soft cooked whole grain (brown rice, buckwheat, barley or millet)
4 Tbsp nut butter (peanut or sesame)
½ cup whole wheat flour and/or soft bread crumbs
1 small onion, finely chopped
1 to 2 Tbsp Tamari soy sauce
Optional seasoning: sea salt, sage, thyme (½ tsp each)

Mix all ingredients well. Shape into ½-inch thick patties. It may be necessary to add a little more flour or water to make mix hold together. Dust patties with flour. Fry in a little oil on low to moderate heat until golden outside and heated through. Serve on homemade buns with all the fixings.

Note: Mung bean sprouts may be substituted for part of the sprouts for a surprising fresh flavor.

Green Rice

Serves 6

3 cups hot cooked brown rice
1 cup green pepper dices
¼ cup Tamari soy sauce
¼ cup toasted sesame seeds
3 Tbsp unrefined oil

Mix all ingredients. Serve with additional herbs to taste.

Rice with Cheese

Serves 4

2 cups cooked brown rice
3 Tbsp unrefined oil
1 onion, chopped
1 clove garlic, minced
2 cups milk
1 cup grated cheese
3 Tbsp chopped parsley

Saute onion and garlic in oil. Add milk, rice, and half the cheese. Mix well, turn into oiled casserole dish. Sprinkle remaining cheese on top. Bake 30 minutes at 350°F. Garnish with parsley.

Basic Soybeans

Serves 8

2 cups dry soybeans
6 cups water

Soak overnight. Next day, add more water if needed. Cover and simmer 2 to 3 hours or until tender. Do not season until after soybeans are tender.

Basic Whole Grains
Serves 8

4 cups seasoned stock or
3½ cups water & ½ cup tamari soy sauce
2 cups whole grain: wheat, rye, bulgur, or barley
2 Tbsp unrefined oil
¼ cup onion dices

Boil stock in top of double-boiler over direct heat. Add remaining ingredients. Cover and set over bottom of double-boiler containing hot water. Cook over low heat until liquid is absorbed. Season to taste.

Alternate method: Bring stock to boil. Add grain. Cover and simmer over low heat until liquid is absorbed, 1 to 1½ hours. Add remaining ingredients.

Rice, Wheat or Rye Flakes Casserole
Serves 4-6

1 Tbsp unrefined oil
2 cups rice, wheat or rye flakes
4 cups water
½ tsp sea salt
1 Tbsp unrefined oil
1 cup onion dices
2 cups mung bean sprouts
¼ cup cheese powder

In deep saucepan, saute flakes in oil until they begin to brown. Add water and salt. Cover and simmer until liquid is absorbed. Add remaining ingredients. Place in oiled casserole dish and bake at 350°F. just until heated through.

Molasses Pintos
Serves 8

2 cups dry pinto beans
5 cups water
2 small onions, chopped
1 Tbsp chili powder
1 cup stewed tomatoes, or tomato sauce
¼ cup molasses
¼ cup Tamari soy sauce

Add pintos to water. Cover and simmer until tender, 2 to 3 hours. Add other ingredients and simmer until tender.

Potato Pancakes

Yield: 6 cakes

1 onion, diced
½ tsp sea salt
2 cups cold mashed potatoes
  or reconstituted potato flakes
2 Tbsp parsley
Whole wheat flour
Optional: 1 egg, beaten

Mix onion, salt, potatoes and egg (if desired). Add parsley and enough flour to hold them together. Fry in oil until brown on one side, then turn and brown opposite side.

Stoved Potatoes

2 cups potato flakes
3 cups water
½ tsp sea salt or vegetable
  salt (herb)
2 Tbsp butter powder

Bring water to boil. Remove from heat. Add salt. Stir in potato flakes and butter powder. Using herbs or herb salt gives this dish a very interesting taste. Let set 5 minutes until mixture thickens.

Crunchy Rice

Serves 4

2 cups cooked rice
1 cup cheese powder
½ tsp sea salt
2 cups milk
3 eggs, beaten
1 onion, chopped
½ cup slivered almonds
¼ cup sunflower seeds
½ cup onion dices

Combine all ingredients. Turn into oiled casserole dish. Bake 35 minutes at 350°F.

Huevos Rancheros on Rice

Serves 6

2 cups water
4 tomatoes, chopped
1 cup raw brown rice
2 Tbsp unrefined oil
1 onion, chopped
1 green pepper, chopped
1 tsp sea salt
1 tsp paprika
6 eggs

Bring water and tomatoes to a boil in large, deep skillet. Add the rice. Cover and simmer about 40 minutes. Saute onions and peppers in oil until soft. Add to cooked rice. Add salt and paprika, mix thoroughly. Make 6 indentations in rice. Drop one egg into each indentation. Cover and continue to simmer until eggs are set (5 to 10 minutes). Serve with chili and tomato sauce if desired.

Scrambled Rice

Serves 6-8

3 Tbsp unrefined oil
1 cup onion or green pepper dices
4 cups soft cooked brown rice
2 cups scrambled eggs
1 tsp sea salt
Optional: mushrooms

Heat large, heavy skillet. Add oil and rice and saute 5 minutes. Add reconstituted freeze-dried eggs and the salt and heat, stirring frequently until mix is hot. Add chives, pepper or other seasonings as desired. Add leftover vegetables and heat through.

Lentil and Tomato Loaf

Serves 6

1½ cup water or stock
2 cups cooked lentils
2 Tbsp unrefined oil
½ cup onion dices
½ cup celery dices
1 cup tomato granules
¼ cup wheat germ
Sea salt to taste

Mix all except wheat germ and salt. Simmer until thick. Mix in wheat germ and salt. Stir in a little whole wheat flour if needed to thicken. Bake in oiled loaf pan 30 minutes at 350°F. Any other cooked and mashed beans or grains may be substituted for lentils. Other vegetable dices may be added, or tuna or beef dices for a change of taste.

Rice Patties

Serves 4

2 cups cooked brown rice
¼ cup unrefined oil
1 egg, beaten
1 tsp sea salt
¼ cup soy flour
¼ cup whole wheat flour

Mix all ingredients into soft dough. Shape into patties. Fry in hot oil. Add Tamari to taste.

Soy Patties

Serves 6

2 cups cooked soy flakes
4 Tbsp unrefined oil
½ cup corn germ
½ cup triticale or other flour
2 eggs, beaten
2 Tbsp Tamari soy sauce

Mix all ingredients, adding flour or water to make patties stick together. Shape and fry in hot oil.

Barley Rice Burgers
Serves 10

3 cups cooked brown rice
2 cups cooked barley
1 cup yellow cornmeal
½ cup whole-wheat flour
1 cup onion dices
Seasoning to taste

Mix all ingredients in a large bowl. Add water until mixture forms a tight ball in your hand. Form into circular flat patties and pan fry until golden crisp. (If a lighter mixture is desired, add a bit of moisture.)

BREADS

Apple Muffins
Yield: 12 large

1 cup raw honey
1 cup unrefined oil
4 eggs
1 tsp vanilla
2½ cups whole wheat flour
1 tsp low-sodium baking powder
½ tsp sea salt
1 tsp each allspice, nutmeg, cinnamon
½ cup non-instant dry milk powder
2 cups apple dices
1 cup chopped nuts

Beat the honey, oil, eggs, and vanilla. Mix all dry ingredients. Add to honey mixture. Stir well. Fold in apples and nuts. Bake in muffin tins at 400°F for 12 to 15 minutes. Remove from tins and cool.

Cornmeal Dumplings

2 cups cornmeal
1 tsp sea salt
1 egg, beaten
Boiling water
¼ cup rice flour

Mix cornmeal and salt. Stir in egg. Pour enough water over mixture to make a thick paste. Stir thoroughly and form into small balls. Roll in rice flour. Drop dumplings into simmering soup or stew. Cover and continue to cook for 10 to 15 minutes.

Basic Muffins

Yield: 10 muffins

1 cup mixed whole grain flours
½ tsp sea salt
1 tsp (heaped) low-sodium baking powder
1 Tbsp unrefined oil
1 egg, beaten
1 Tbsp dark molasses
1 cup buttermilk

Mix dry ingredients. Stir in oil and rub in well with fingers. Add egg, molasses and buttermilk. Stir well. Fill greased muffin tins a little over half-full. Bake 15 to 17 minutes at 450°F. These muffins are excellent with ½ cup blueberries or other small berries stirred into batter. Wheat germ may be used as part of the flour.

Elmer's Cornpone

2 loaves, 5″ x 9″ pan

2 cups milk
1 cup water
3 Tbsp honey
2 Tbsp dry yeast
2 tsp sea salt
1 cup yellow cornmeal
6 cups whole wheat flour

Mix liquids together in saucepan and heat until lukewarm. Add yeast, salt, and flours. Mix and let set for 10 minutes. Knead into the dough 1 more cup of flour, working 5 minutes. Place in an oiled bowl, cover and let rise to double in size. Knead again and shape into 2 loaves. Bake at 350° for 35 minutes. Turn out onto rack to cool, and brush crusts with oil. Cover with towel while cooling.

Fast Whole Wheat Yeast Bread

1 loaf, 5″ x 9″ pan

13 ounces lukewarm water
1 Tbsp dry yeast
1 tsp raw honey
Optional: 2 Tbsp unrefined corn germ oil or safflower oil
4 cups whole wheat flour
1 tsp sea salt

Mix honey in warm water. Sprinkle yeast on top of water. Set aside until frothy, about 5 minutes. Add oil, if desired, to yeast mixture. Mix flour and salt. Pour yeast mixture into center of the flour. Mix thoroughly, by hand if necessary. Shape dough into a loaf. Place into an oiled loaf pan. Let rise slightly over the top of pan, 20 to 30 minutes. Bake 30 to 40 minutes at 350°F.

The oil in this recipe is strictly optional. The loaf is delicious either way. It can be made from start to finish in one hour, including baking. Actual mixing time is only 5 to 10 minutes.

Chewy Batter Bread
Serves 4

1 egg	½ cup cooked brown rice
1 tsp sea salt	½ cup cornmeal
Boiling water	

Beat egg until light. Add salt, rice, and cornmeal. Mix well. Stir in boiling water until mixture becomes as thick as heavy cream. Pour into oiled 8 x 8 inch pan. Bake 20 to 30 minutes at 350°F. This recipe makes a crisp, chewy bread that is good with soup. It is a good way to use leftover rice.

Unleavened Bread
Yield: 1 loaf

1⅔ cups warm water	4 cups whole wheat flour

Pour water into large bowl. Add 3 cups flour. Stir. Add last cup of flour gradually, stirring until the batter is stiff. Knead in remaining flour. Knead about 20 times, until water and flour are well mixed. Shape into loaf. Place the loaf into a small oiled loaf pan. Cover with a damp towel and let rise in a warm place for 24 hours. In the summer rising time is less, depending on your weather temperature. The towel should be kept damp throughout the rising time. Bake 1 hour at 350°F. The crust will be chewy and hard. If bread did not rise to the top of the loaf pan, bake 30 minutes at 250°F. and then 350°F. until crust is hard. Cool on rack.

Flour Tortillas
Yield: 2 dozen

4 cups whole wheat pastry flour	⅓ cup unrefined oil
1 tsp sea salt	1 cup warm water

Mix flour and salt. Add oil and mix together with fingertips. Stir in enough water to make a firm ball of dough. Knead until smooth. Let set 20 minutes. Pinch off piece of dough the size of golf ball. Roll it out on a floured board to a more or less round shape about 4 inches in diameter. Cook on an unoiled hot griddle or in a skillet, about 2 minutes on each side. You can salt the griddle to keep the tortilla from sticking, but it is not really necessary.

Tea Pancakes or Crumpets

2 eggs
2 Tbsp honey
2 cups whole wheat flour
½ tsp salt
2 Tbsp oil or butter
1½ cups milk or buttermilk

Beat eggs well. Add honey. Mix dry ingredients and add alternately with oil. Slowly add enough milk so batter is consistency of thick cream. Oil hot griddle, beat batter well and drop by large tablespoonfuls, spreading as thinly and evenly as possible. When golden brown, turn and cook on other side. Lay pancakes on a clean towel and spread with butter, jam or honey. Roll up and serve warm.

Cornbread

2 eggs, separated
1½ cups water or other liquid
¼ cup unrefined oil
1 cup cornmeal
1 cup pastry or whole wheat flour
2 tsp low-sodium baking powder
Dash of sea salt

Beat egg yolks. Add liquids. Mix in dry ingredients. Beat egg whites until stiff and fold in. Bake at 400°F. for 25 minutes in preheated and oiled cast-iron pan.

Indian Spoonbread

Serves 6-8

1 cup cornmeal
1 tsp salt
2 cups boiling water
1 cup cold milk
4 eggs
5 Tbsp butter, melted

Combine cornmeal and salt in mixing bowl. Stir in boiling water till smooth. Let stand for few minutes, then mix in cold milk. Add one egg at a time, beating hard after each addition. Stir in melted butter.

Butter a medium-sized earthenware or glass baking dish. Pour batter into baking dish and bake at 425°F. for 30 minutes. Serve hot with butter.

Swedish Rye

Yield: 3 round loaves

3 cups warm water
2 Tbsp yeast
½ cup honey
½ cup molasses
6 cups (approx.) whole wheat flour
½ cup oil
1 Tbsp salt
4 cups rye flour
Sprinkle of corn meal

Combine water, yeast, honey and molasses. Let set about 5 minutes until yeast bubbles to the surface. Add 3 cups whole wheat flour and stir thoroughly. Let rise 15 to 30 minutes. Stir down. Add the oil, salt, rye flour, and enough whole wheat flour to make a fairly stiff dough. Knead thoroughly. Let rise until doubled in bulk, usually about one hour. Punch down. Shape into three round loaves. Place on cookie sheet sprinkled with cornmeal. Cut a slit in the top of each. Let rise until almost double, about ½ hour. Bake at 375°F. for 40 minutes or until done. Cool on rack.

DESSERTS

Granola Candy

Yield: 2 dozen small balls

½ cup old-fashioned peanut butter
½ cup raw honey
½ cup granola or crushed wheat flakes
¼ cup non-instant dry milk powder

Combine peanut butter with honey. Add granola and milk powder. Shape into a long roll. Chill until firm. Slice or roll into balls. Add milk powder if needed to stiffen.

Date Balls

1 cup date sugar
1 cup milk powder
½ cup raisins
1½ cup chopped almonds

Mix ingredients. Add warm water to make thick, stiff mix. Shape into small balls and roll in grated coconut or finely chopped nuts. Store in airtight container.

Peanut Butter Balls

½ cup raw honey
½ cup old-fashioned peanut butter
½ cup corn germ
¾ cup milk powder

Mix all ingredients well. Roll into small balls. Roll in ground nuts if desired. Chill.

Quick Cobbler

Serves 6

6 Tbsp butter
1½ cups whole wheat flour
¼ tsp sea salt
¼ tsp low-sodium baking powder
½ cup raw honey
1 egg, beaten
1 cup milk
1 cups fresh fruit (with ¼ cup more raw honey if fruit is tart)

Melt butter in 1½-quart flat baking dish. Mix flour, salt, baking powder, honey, egg, and milk. Pour in the center of the melted butter in the dish. Pour fruit over top. Bake 30 to 35 minutes at 400°F.

Apple Crisp

Serves 4-6

1 pound sliced apples
1 cup whole wheat pastry flour
½ cup unrefined oil
¼ cup honey
1 cup raisins

Arrange a layer of sliced apples in bottom of well-oiled baking dish. Mix remaining ingredients and cover apples with a layer of mixture. Repeat layers. Bake until golden and sizzling, about 40 minutes at 350°F.

Wheat Sprout Pudding

Serves 6

1 cup wheat sprouts
½ cup whole wheat flour
½ cup date sugar
1 cup sunflower seeds

Mix all ingredients. Add enough water to make batter the consistency of thick cream. Pour into oiled pan. Bake about 1 hour at 325°F. Serve hot or cold. Good topped with fresh fruit.

Oatmeal Cookies

Makes 3 dozen

½ cup honey
1 Tbsp unrefined oil
2 eggs, beaten
Grated rind of one lemon
½ tsp sea salt
2 cups oatmeal
½ cup whole wheat flour

Mix all ingredients together to make a stiff dough. If too thick, add a little milk. Drop by teaspoonfuls onto oiled cookie sheet. Bake 8 to 10 minutes at 400°F.

Backpack Snack

Wheat or rye flakes
Sesame seeds
Sunflower seeds
Whole or chopped nuts
Raisins
Dried fruits, chopped

Toast flakes, sesame seeds, and sunflower seeds in ungreased skillet. Pour into bowl and mix with nutmeats, raisins, and any desired chopped dried fruits. Cool mixture completely. Carry in tightly closed plastic bags or other airtight containers.

An Annotated Bibliography on Food and Preparedness

NOTE: The following bibliography by Dr. Earl G. Alexander is excerpted from THE SIMPLER LIFE MENU PLANNER, by Charlotte Duvall Savage and Earl Alexander, © 1982, Harvest Press, P.O. Box 7971, Waco, Texas 76710 and is used with permission. The book is a primer for menu planning, containing 280 recipes, defining maximum use of the natural foods in The Simpler Life Food Reserve program.

Getting acquainted with the following information sources will aid you in learning more about good foods and preparedness. Although the following list of books is substantial, it is only a sampling of those available on the subject of whole natural foods. Asterisks (*) are used to designate "The Essential Library" for Natural Foods Cooking and Food Preparedness.

NUTRITION TEXTS, COMMENTARIES

The Complete Shopper's Guide to Natural Foods, Christopher Kilham, 1980, Autumn Press, 1318 Beacon St., Brookline, MA 02146. Includes information on vitamins, supplements, cosmetics, kitchenware, and body care tools.

Christian Health Newsletter, Lester C. and Marjorie H. Pridgen, a newsletter published monthly by Maranatha Foundation, 7061 Old Kings Rd., S. 268, Jacksonville, FL 32217. Described by the editors as "a non-medical publication about alternate methods of healing by God's natural and supernatural power."

Consumer Beware!, (Your Food and What's Been Done to It), Beatrice Trum Hunter, 1971, Simon and Schuster, New York, NY 10020. Cites what can be done to halt and reverse the shocking deterioration of our basic foods: Production, processing, packaging, labeling, and distribution.

Dictionary of Nutrition, Richard Ashley and Heidi Duggal, 1976, Pocket Books, New York, NY 10018. Alphabetical format food facts, including nutritional properties.

A Diet for Living, Jean Mayer, 1975, Book Department, Consumers Union, Orangeburg, NY 10962. Nutritional facts by one of the nation's foremost authorities.

* *Diet for a Small Planet,* Frances Moore Lappe, Revised Edition 1976, Ballantine Books, New York, NY 10022. Principles of high-protein meatless cookery. Includes 58-page summary of the world food situation.

The Good Seeds, The Rich Grains, The Hardy Nuts for a Healthier, Happier Life, Ruth Adams and Frank Murray, 1973, Larchmont Books, New York, NY 10036.

The Healing Power of Natural Foods, May Bethel, 1978, Wilshire Book Company, 12015 Sherman Road, No. Hollywood, CA 91605. Information helpful in preventing disease and building health.

Improving Your Child's Behavior Chemistry, Lendon H. Smith, M.D. 1976, Pocket Books, New York, NY 10020. Eating habits and vitamin imbalances that may be making your child unruly, unresponsive, and unhappy.

Let's Eat Right to Keep Fit, Adelle Davis, 1954, 1970, Signet Books, New York, NY 10019. A guide to physical and emotional well-being through proper diet.

* *Let's Try Real Food,* Ethel H. Renwick, 1976, Zondervan, Grand Rapids, MI 49506. Good nutrition and health, approached from a perspective of Christian stewardship.

The Love Feast, Graham Kerr, 1978, Simon and Schuster, New York, NY 10020. Wholesome food helps create wholesome Christian family life. Recipes and special notes with nutrition profiles from the author of *The Galloping Gourmet* and *The New Seasoning.*

The Miracle Nutrient, Carl I. Flath, 1975, Bantam Books, New York, NY 10017. Importance of natural fiber in daily diet.

* *The Natural Food Primer,* Beatrice Trum Hunter, 1972, Simon and Schuster, New York, NY 10020. Help for the beginner from the "first lady" of natural foods.

Nutrition Against Disease, Roger J. Williams, Ph.D., 1973, Bantam Books, New York, NY 10019. How proper eating is preventive medicine.

Psychodietetics, E. Cheraskin, W.M. Ringsdorf, Jr. with Arline Brecher, 1974, Bantam, New York, NY 10017. Two doctors show how nutrition affects your personality, mood, and emotions.

* *Sugar Blues,* William Dufty, 1975, Warner Books, New York, NY 10019. Exposing sugar, the killer in your diet. Offering a sugar-free way to health.

* *Total Nutrition During Pregnancy (The Kamen Plan),* Betty Kamen and Si Kamen, 1981, Appleton-Century-Crofts, 292 Madison Ave., New York,

NY 10017. Well-documented. Presents safe, sane, simple regimen of nutritional guidelines pregnant women can follow with confidence, preventing health problems for mother and child.

Tracking Down Hidden Food Allergy, William G. Crook, M.D., 1980, Professional Books, P.O. Box 3494, Jackson, TN 38301. Cleverly illustrated book for children and adults; easy to follow instructions that help you carry out elimination diets; will help identify whether symptoms are caused by adverse or allergic reactions to foods in daily diets.

RECIPES, COOKBOOKS

The Back to Eden Cookbook, Jethro Kloss, 1974, Woodbridge Press, Santa Barbara, CA 93111. More than 260 recipes and helpful information on natural foods for health and healing from the author of *Back to Eden.*

The Beginner's Natural Food Guide and Cookbook, Judy Goeltz, 1978, Hawkes Publishing, Salt Lake City, UT 84115. A what, why, why not, and how book including menus and recipes.

* *The Complete Sprouting Cookbook,* Karen Cross Whyte, 1973, Troubador Press, San Francisco, CA. What to sprout, how to do it, and how to use sprouts in delicious ways.

* *The Deaf Smith Country Cookbook: Natural Foods for Family Kitchens,* Marjorie Winn Ford, Susan Hillyard, and Mary Faulk Koock, 1973, Collier Books, MacMillan, New York, NY 10022. (Available for $6.95 prepaid from Arrowhead Mills, Inc., P.O. Box 866, Hereford, TX 79045.) Recipes for whole foods cooking.

From God's Natural Storehouse, Yvonne G. Baker, 1980, David C. Cook Publishing Co., Elgin, IL 60120. Practical alternatives to cooking with junk.

Future Food, Colin Tudge, 1980, Harmony Books, New York, NY. A very readable book by an Englishman. Politics, philosophy, and recipes for the 21st century.

* *The Good Breakfast Book,* Nikki and David Goldbeck, 1976, Links Books (Div. of Music Sales Corp.), New York, NY 10023. A complete nutrition guide and 400-plus recipes for the most important meal of the day. (Foreward by Beatrice Trum Hunter.)

A Guide for Nutra Lunches and Natural Foods, Sara Sloan, 1977, Sara Sloan, NUTRA, P.O. Box 13825, Atlanta, GA 30324. How to get nutrition on your school lunch menu. Great ideas for kids and their parents.

Laurel's Kitchen, Laurel Robertson, Carol Flinders, and Bronwen Godfrey, 1976, Bantam Books, New York, NY. A handbook for vegetarian cookery and nutrition.

* *The Living Cookbook,* by Yvonne Turnbull, 1981, Omega Publications, Box 4130, Medford, OR 97501. Star of 700 Club's weekly kitchen

show, Mrs. Turnbull provides instructions for making easily prepared, nutritious dishes from scratch, using natural and basic ingredients.

* *The Natural Foods Cookbook,* Beatrice Trum Hunter, 1961, Fireside Book by Simon and Schuster, New York, NY 10020. The first book of Natural Foods recipes (over 2000).

Old Fashioned Recipe Book, Carla Emery, 1977, Bantam, New York, NY 10019. Interesting book with much on homesteading, recipes, etc. There's far too much sugar in many recipes, however.

Recipes for a Small Planet, Ellen Buchman Ewald, 1973, 1976, Ballantine Books, New York, NY 10022. The art and science of high-protein vegetarian cookery.

Recipes for Life, Dr. Ann Wigmore, 1978, Rising Star Publications, 25 Exeter Street, Boston, MA 02116. A book of completely raw food recipes.

* *The Simpler Life Cookbook from Arrowhead Mills,* Frank Ford, 1974, Harvest Press, P.O. Box 7971, Waco, TX 76710. (Also available from Arrowhead Mills, Inc., P.O. Box 866, Hereford, TX 79045—Three copies prepaid for $2.00.) Recipes, nutrition, and preparedness.

* *Snackers,* Maureen and Jim Wallace, 1978, Madrona Publishers, Seattle, WA. 230 recipes for tasty, nutritious, healthful, easy-to-prepare snacks. Helps kick the junk food habit.

Tofu Goes West, Gary Langrebe, 1978, Fresh Press, Palo Alto, CA 94303. Tofu in Western trappings. Perfect fixin's for our time using tofu, a natural foods protein staple in the Orient for centuries.

The Vegetarian Family, Runa and Victor Zurbel, 1978, Prentice-Hall, Englewood Cliffs, NJ Practical advice on high-protein vegetarian meals.

FOOD COMPOSITION TABLES

Composition of Foods, Raw, Processed, Prepared, U.S.D.A. Agriculture Handbook No. 8, Reprinted 1975, Stock 001-000-00768-8, Superintendent of Documents, U.S. Government Printing Office, Washington, DC 20402. Every dietitian should know his or her way around in this set of tables.

Nutritive Value of Foods, U.S.D.A. Home & Garden Bulletin 72, Revised 1977, Stock 001-000-03667-0, same address as above.

FARMING, GARDENING

Acres, U.S.A., monthly publication, Acres U.S.A., P.O. Box 9547, Raytown, MO 64133. Available by subscription. A voice for eco-agriculture. "To be economical, agriculture has to be ecological."

Country Christian, published bi-monthly, Route 2, Box 311-C, Santa Fe,

NM 87501. Helpful information both for the home steader and would-be farmer.

Encyclopedia of Organic Gardening, edited by J.I. Rodale and Staff, 1975, Rodale Books, Emmaus, PA 18049.

Growing Up Green, Alice Skelsey and Gloria Huckaby, 1973, Workman Publishing Co., New York, NY 10022. Highly acclaimed project book about the marvels of nature, parents and children gardening together.

* *Grow It!,* Richard W. Langer, 1972, Avon Books, The Hearst Corp., New York, NY 10019. The beginner's complete in-harmony-with-nature small farm guide from vegetable and grain-growing to livestock care.

How YOU Can Grow ALL of Your Own Food (A Natural Step-By-Step Method), William Behr Mueller, 1976, Big Toad Press, 617 25th St., Sacramento, CA 95816. Incremental steps toward food self-sufficiency; includes recipes helpful to vegetarians.

More Food from Your Garden, J.R. Mittleider, 1975, Woodbridge Press, Santa Barbara, CA 93111. How to make "grow boxes," custom-made soil. Includes techniques of hydroponic, organic, and conventional gardening methods. Many helpful ideas, but go easy on the chemicals!

The Mother Earth News, a magazine published bi-monthly, P.O. Box 70, Hendersonville, NC 28791. Much more than farming, gardening, and canning information.

Organic Gardening, monthly publication by Rodale Press, Emmaus, PA 18049, available by subscription.

The Organic Way to Plant Protection, 1966, Rodale Press, Emmaus, Pennsylvania 18049. A complete garden reference on controlling insects and plant diseases without DDT and other pesticides, by the editors of *Organic Gardening.*

PREPAREDNESS, PRESERVATION & STORAGE OF FOODS

Boy Scout Handbook, Eighth Edition, 1975, Boy Scouts of America, North Brunswick, NJ Good information on hiking, camping, cooking, environment, survival.

Blue-Print for Survival, by the editors of *The Ecologist,* 1974, Signet Books, New York, NY 10019. Addresses the population/environment/resource dilemma. Has interesting information, but many answers are far too humanistic!

Eat, Drink, and Be Ready, Monte L. Kline and W.P. Strube, Jr., 1977, Harvest Press, P.O. Box 7971, Waco, TX 76710. Life in the last days, a philosophy of preparedness, and a look at food, nutrition, and disease.

Emergency/Survival Handbook, published by American Outdoor Safety

League, available from The Mountaineers Books, 719-B Pike St., Seattle, WA 98101. Quick, indexed basic information on how to react to and cope with emergencies and survive the unexpected. Includes lists of basic equipment and first-aid and survival-kit supplies.

End Times News Digest, Jim McKeever, a newsletter published by Living Waters Ministries, c/o Alpha Omega Publishing Company, P.O. Box 4130, Medford, OR 97501. Includes a physical preparation section which deals with various aspects of a self-supporting life-style.

Famine and Survival in America, Howard J. Ruff, 1974, Abridged version 1978, Target Publishing, P.O. Box 172, Alamo, CA 94507.

Life After Doomsday, Bruce D. Clayton, 1980, Paladin Press, P.O. Box 1307, Boulder, CO 80306. A survivalist guide to nuclear war and other major disasters.

* *Making the Best of Basics: Family Preparedness Handbook,* James Talmage Stevens, 1977, Peton Corporation, P.O. Box 11925, Salt Lake City, UT 84111. Many excellent preparation charts and aids. Many of the recipes have sugar in them, though.

Natural Foods Storage Bible, Sharon B. Dienstbier and Sybil D. Hendricks, 1976, Horizon Publishers, Bountiful, UT 84010. Secrets of storing, cooking, and enjoying natural foods.

* *Occupy: An End-Time Strategy for the Believer,* Frank Ford, 1979, Harvest Press, Waco, TX 76710. (Also available from Arrowhead Mills, Inc., P.O. Box 866, Hereford, TX 79045—Five copies prepaid for $2.00.) It's not a question of how much food there is, but whether it's in the right place at the right time. *Occupy* asks the question, "Does God want you to make provision for a time of shortage?" Only you and He can decide.

The Omega Generation, Nate Krupp, 1977, New Leaf Press, Harrison, AR 72601. The world and the church in transition.

Outback with Jesus, Bob Summers, 1975, Harvest Press, Waco, TX 76710. Survival Christianity, a message to pilgrims living in the twilight of time.

Passport to Survival: Four Foods and More to Use and Store, Esther Dickey, 1969, Bookcraft, Inc., Salt Lake City, UT.

The Po'r Boy's Guide to Survival, Bruce Perron, 1981, Cen-Tex Sales Associates, P.O. Box 757, Hewitt, TX 76643. Practical survey of methods and survival techniques for the non-affluent, working person; with addenda and sources of supply.

Remnant Review, Gary North, a newsletter published bi-weekly. Available by subscription, P.O. Box 39800, Phoenix, AZ 85069. An explicitly Christian newsletter in the spirit of Matthew 6:33-34 and pro-free-market in perspective. Attempts to apply biblical principles to economic analysis.

The Standard First Aid & Personal Safety, prepared by the American Red Cross, Second edition, 1979, Doubleday & Co., Garden City, NY. Used for the instruction of first-aid classes.

Stocking Up: How to Preserve the Foods You Grow, Naturally, by the editor of *Organic Gardening and Farming*, 1973, Rodale Press, Emmaus, PA 10849.

Survival Food Storage, Mark and Zhana Thomason, 1980, TSI Publishers, P.O. Box 22009, San Francisco, CA 94122. Imaginative ways to store a year's supply of food for every family member in one or two rooms.

A Survival Guide for Tough Times, Mike Phillips, 1979, Bethany Fellowship, Inc., 6820 Auto Club Road, Minneapolis, MN 55438. How to have mental, economic, physical, and spiritual health for the days ahead.

The Survival Handbook, Bill Merrill, 1974, Arco Publishing Co., Inc., 219 Park Ave. So., New York, NY 10003. Used extensively in earlier editions by U.S. armed forces for survival training.

Survive, a quarterly magazine by Survive Publications, 5735 Arapahoe Ave., Boulder, CO 80303. Regular information explaining potential or threatened problems (economic depression or collapse, natural disasters, diminishing natural resources, civil insurrection, and nuclear attack) and offering solutions or alternatives.

ECOLOGY, STEWARDSHIP OF THE EARTH

Farming and Gardening in the Bible, Alastair I. MacKay, 1950, 1970, Rodale Press/Pyramid Books, New York, NY 10022. The inspiring and illuminating story of the vital role of conservation in biblical times.

Hard Tomatoes, Hard Times, Jim Hightower, 1978, Schenkman Publishing Co., Inc., Cambridge, MA 02138. The original Hightower report, unexpurgated, of "the Agribusiness Accountability Project on the Failure of America's Land Grant College Complex and Selected Additional Views of the Problems and Prospects of American Agriculture in the Late Seventies."

Living Earth, Peter Farb, 1959, Harper Colophon Books, New York, NY 10022. Caring for the soil.

New Harvest, Frank Ford, 1974, Harvest Press, Waco, TX 76710. Poetry, photographs, Christian witness, harmony with the earth.

Our Margin of Life, Eugene M. Poirot, 1978, *Acres, U.S.A.*, P.O. Box 9547, Raytown, MO 64133. Poetry in prose interrelating man, animals, plants, microbes, sun, soil, air, and water. The story of a prairie farm.

The Pesticide Conspiracy, Robert Van Den Bosch, 1978, Doubleday, Garden City, NY An alarming look at pest control and the people who keep us "hooked" on deadly chemicals.

SHARING FOOD

Food First: Beyond the Myth of Scarcity, Frances Moore Lappe and Joseph Collins with Cary Fowler, Houghton Mifflin, 1977, available from Institute for Food and Development Policy, 2588 Mission Street, San Francisco, CA 94110.

Home Missions magazine, November/December 1975 special issue on Hunger, Southern Baptist Convention Home Missions Board, 1350 Spring St., N.W., Atlanta, GA 30309.

Rich Christians in an Age of Hunger, Ronald J. Sider, 1977, Intervarsity Press, Downers Grove, IL 60515. A Biblical study: Poor Lazarus and rich Christians. A Biblical perspective on the poor and possessions. Implementation of simpler life-style.

What Do You Say to a Hungry World?, W. Stanley Mooneyham, 1975, Word Books, Waco, TX. Christian response to the world food problem.

World Vision magazine, a monthly magazine sent to those who contribute to World Vision International, a Christian relief organization, 919 West Huntington Dr., Monrovia, CA 91016.

WORLD FOOD RESOURCES AND PROBLEMS

Building a Sustainable Society, Lester R. Brown, 1981, Norton Publishers, available from Worldwatch Institute, 1776 Massachusetts Ave. N.W., Washington, DC 20036. Addresses problems of eroding cropland, deforestation, oceanic pollution, dwindling resources, food shortages, overpopulation, nuclear proliferation. Suggests a program of efforts for turning the tide.

"Can the World Feed its People?", by Thomas Y. Canby, pp. 2-31 in the July 1975 issue of *National Geographic,* 17th and M Streets, N.W., Washington, DC 20036. An excellent review article.

Food for People, Not for Profit, edited by Catherine Lerze and Michael Jacobson, 1975, Ballantine Books, New York, NY 10022. A sourcebook on the food crisis.

Harvesting the Earth, Georg Borgstrom, 1973, Abelard-Schuman, An Intext Publisher, New York, London. World Food Resources, Agricultural Economics.

The Nutrition Factor: Its Role in National Development, Alan Berg, 1973, The Brookings Institution, 1775 Massachusetts Ave., N.W., Washington, DC 20036. Report of a study sponsored jointly by the Foundation for Child Development and the Brookings Institution.

"U.S. Agriculture is Growing Trouble as Well as Crops," by Wilson Clark, pp. 59-65, in the January 1975 issue of *Smithsonian* magazine, 900 Jefferson Dr., Washington, DC 20560. Overdependence on energy and machines.